The Last Fish Swimming

Recent Titles in Global Crime and Justice

Outsmarting the Terrorists
Ronald V. Clarke and Graeme R. Newman

Human Trafficking, Human Misery: The Global Trade in Human Beings
Alexis A. Aronowitz

Police Use of Force: A Global Perspective
Joseph B. Kuhns and Johannes Knutsson, Editors

Beyond Suppression: Global Perspectives on Youth Violence
Joan Serra Hoffman, Lyndee Knox, and Robert Cohen, Editors

Identity Theft in Today's World
Megan McNally

Sold into Extinction: The Global Trade in Endangered Species
Jacqueline L. Schneider

Corporal Punishment around the World
Matthew Pate and Laurie A. Gould

Internet Child Pornography: Causes, Investigation, and Prevention
Richard Wortley and Stephen Smallbone

Dying with Dignity: A Legal Approach to Assisted Death
Giza Lopes

The Last Fish Swimming

The Global Crime of Illegal Fishing

GOHAR A. PETROSSIAN

Global Crime and Justice
Graeme R. Newman, Series Editor

PRAEGER™

An Imprint of ABC-CLIO, LLC
Santa Barbara, California • Denver, Colorado

Library of Congress Cataloging-in-Publication Data

Names: Petrossian, Gohar A., author.
Title: The last fish swimming: the global crime of illegal fishing / Gohar A. Petrossian.
Description: Santa Barbara, California : Praeger, an imprint of ABC-CLIO, LLC, [2019] | Series: Global crime and justice | Includes bibliographical references and index.
Identifiers: LCCN 2018061598 (print) | LCCN 2019001581 (ebook) | ISBN 9781440830426 (ebook) | ISBN 9781440830419 (hard copy : alk. paper)
Subjects: LCSH: Fishery management, International—Law and legislation. | Fisheries—Corrupt practices.
Classification: LCC K3897 (ebook) | LCC K3897 .P48 2019 (print) | DDC 346.04/6956—dc23
LC record available at https://lccn.loc.gov/2018061598

ISBN: 978–1–4408–3041–9 (print)
978–1–4408–3042–6 (ebook)

23 22 21 20 19 1 2 3 4 5

This book is also available as an eBook.

Praeger
An Imprint of ABC-CLIO, LLC

ABC-CLIO, LLC
147 Castilian Drive
Santa Barbara, California 93117
www.abc-clio.com

This book is printed on acid-free paper ∞

Manufactured in the United States of America

Contents

Illustrations

Figures

Tables

Boxes

Series Foreword

It seems like the problem of climate change as the fault of humans pales in comparison to the global human gluttony of fish. The extent of damage, destruction, willful consumption, and harvesting of fisheries all over the world bears this out. And we are the enablers—all of us at some time have eaten an illegal fish or a piece of it. The consumers are the enablers, but the harvesters and massive trawlers clearly feed our gluttony. After all, gluttony is not possible if there is nothing to gluttonize. The demand fuels the catch, but the catch also enables the demand.

This book begins with a sobering account of the gradual (or in some places rapid) destruction of the world's fishing resources. Worse, it reveals that the apparently unstoppable depletion is occurring in distinct geographical and geopolitical regions: regions that are less developed economically—most of them former colonies of Western (and now Eastern) powers. In the classic colonial tradition, these resources are exploited and the fish exported to adorn the tables of the "civilized" world of the East and the West. And in accordance with the great traditions of globalization, many international organizations, especially the several UN affiliates and many privately funded NGO projects, have launched programs to counter this crisis; spawned many conferences; passed many binding and nonbinding resolutions and international laws; published many global regulations; implemented control and surveillance measures; identified high-risk areas, shipping ports, and quasi-criminal organizations and businesses; and tracked and prosecuted the illegal offenders.

Unfortunately, after digesting the extensively documented data of fish destruction and recounting the many preventive measures introduced (often ingenious—using albatrosses as spies on fishing vessels, for example), there appears to be no sign of a winning global strategy, unless it can be argued that maybe the best that can be hoped for is to keep the scourge of illegal fishing at its current (unsustainable) level. But as disturbing as this inevitable tilt toward pessimism may be, the author of this book, Gohar A. Petrossian, offers many rays of hope. Adopting the recent criminological perspectives of situational crime prevention and environmental criminology, she manages to lay out a road map for future approaches to prevention: strategies that focus on reducing the rewards of illegal fishing (fines and regulations), increasing the risks of getting caught (modern surveillance technologies), reducing excuses (we only provide what people need: fish for their diets), controlling the tools used for mass fishing enterprises, increasing the costs of big business fishing, identifying alternatives (farming), and holding countries responsible for lax control of illegal fishing that occurs in their waters.

Though Petrossian recounts several important empirical studies of these various attempts to prevent and control illegal fishing, the scientific basis for this enterprise is in its infancy. Compared to the enormous amounts of scientific research and data amassed in the traditional fields of criminology (murder, rape, street crime, etc.), the criminology of the environment, especially illegal fishing and the exploitation of other endangered species, will require a new generation of criminologists to set it on par with the narrow view of crime so typical of nineteenth- and twentieth-century criminology. Not that traditional criminology has a lot to show for its lavish amounts of scientific research. Nor is it altogether clear that the field of environmental criminology as applied to illegal fishing would be helped much by following the narrow view of crime dominating traditional criminology. Nor is it clear that criminalizing a problem will solve it. Instead, it may be that research on illegal fishing, looking much more closely at the process and outcomes of regulation rather than criminalization (i.e., arrest, finding of guilt, and prison), may help us understand when criminalizing a problem helps and when it does not.

The Last Fish Swimming, the tenth and final book in the Global Crime and Justice series, offers a poignant reminder that there is much to be done and much we do not know. The illegal production and consumption of the world resource of fish is one that directly involves every person on this planet; it is truly global in scope. Unlike the vague threat of climate change that, to make it convincing, must be dressed up as though the end of the world were imminent, the catastrophic destruction of the world's fisheries is clearly evident—its human cause being unquestionable. This final destruction of a key source of food and health represents a massive assault on Nature by human nature. This is not a crime against humanity, but of humanity against itself. It may not be possible to reduce the human appetite for fish, but it may be possible—as the author shows in her reviews of the scientific studies of environmental criminology and situational prevention—to make some small inroads at specific locations, while we all sit back and hope that the many international organizations will begin to assess their effectiveness at intervention and to muster the world support necessary to save our fish from extinction.

Graeme R. Newman
Series Editor

Acknowledgments

This book would not have been possible without the support, guidance, and encouragement of many people. I am indebted to my mentor, Dr. Ronald V. Clarke, for everything he has done for me. Thank you for the fateful meeting where you asked me to study illegal fishing—a meeting that sealed the course of my intellectual inquiry for many years to come. Among other reasons, thank you for playing such an integral role in fostering my intellectual growth. Thank you to Dr. Graeme Newman for his valuable feedback on the earlier drafts of the manuscript. If it were not for his suggestion to "unlearn some of the bad habits of academic writing" (email communication, April 15, 2014), I would not have been able to write this book the way it came out. Thank you to Dr. Daniel Pauly for his meticulous review of my manuscript and his prudent advice on how I can improve it. His guidance was, indeed, instrumental.

I wish to express my gratitude to my colleague and mentor, Dr. Michael Maxfield, or "grandpa Maxfiled" as my little Sophia would say, for his unwavering support. Thank you also to my colleagues Dr. Stephen Pires, Dr. William Moreto, Dr. Nerea Marteache, Dr. Daan van Uhm, Dr. Judith Weis, Dr. Mangai Natarajan, and Dr. Julie Viollaz for investing your criminological training and intellectual energy into the study of wildlife crimes. I would like to extend my gratitude to Jessica Gribble, senior acquisitions editor at ABC-CLIO, for her guidance throughout the development of the book. Thank you for cheering me along as I completed the chapters, and for your positive sprit. I am so fortunate to have had this opportunity to work with her. Thank you, also, to Roza Minasyan for embellishing the book with the exquisite drawings of the fishing vessels.

Above all, I wish to thank my husband, Tigran Nanian, for his limitless patience and unconditional love. Thank you for being my source of inspiration and my laboratory of ideas. Thank you for sacrificing your career to help mine go forward. I am forever fortunate to be your wife.

I am thankful to my parents for their love and support, and for encouraging me to pursue knowledge. I would not have been where I am today without the great sacrifices you made and the deprivations you had to endure to raise me.

This book is dedicated to my daughters Sophia Elen Nanian and Ani Emily Nanian, whose existence has made me strive to become the best I can be to, one day, make them proud.

Funding for this work was provided by a grant from the Office for the Advancement of Research at John Jay College of Criminal Justice, and a City University of New York Office of Research Book Completion Award.

Introduction

The crime of illegal fishing has only recently gained international attention as more evidence compiles to attest to the irreversible damage it causes to the marine ecosystem. Over 70 percent of the world's fish stocks are either fully exploited, overexploited, or depleted, and illegal fishing is one of the major reasons behind this problem. Fisheries scientists estimate that the biomass of large predatory fish, such as swordfish, marlin, large tunas, and cod, has dropped 90 percent from its preindustrial levels, leaving a precarious 10 percent to be fished today. Should fisheries resource depletion persist at the current rate, most global fish stocks will have collapsed by 2050, leaving our oceans empty and depriving millions of people of this critical food source. Illegal fishing also leads to the irreparable damage of the coral reefs and seagrass beds, as well as significantly impacts the populations of nontarget species, such as sea turtles, seabirds, and dolphins.

Coastal countries bear the direct consequences of illegal fishing as 90 percent of these activities occur within their exclusive economic zones. Poor coastal countries are particularly affected because these countries have the richest marine resources that are exploited by domestic and foreign fleets, and they have low monitoring, control, and surveillance capacity. The rising global demand and the increasing value of fish and fish products have thus made illegal fishing a lucrative business. Despite the regulatory measures undertaken internationally, regionally, and locally, the problem persists.

It is clear that a lot remains to be done to deal with the problem of illegal fishing. However, until recently, the issue has been seen as a fisheries management problem rather than a criminal act that requires the collaboration and guidance of crime scientists. With this book, we hope to fill this gap and to add to the toolbox of strategies that can be used to stop illegal fishing. The book draws from the theoretical guidance of environmental criminology and crime science, to scrutinize illegal fishing and suggest ways to prevent it.

The book is divided into seven chapters. Chapter 1 begins with a brief historical overview of fishing and the evolution of fishing into a global, multibillion-dollar industry. The chapter chronicles these historical developments while drawing the readers' attention to the role of opportunity in shaping this development. The chapter also provides an overview of the main sectors of the fishing industry, types of fisheries, and the role of fishers in the industry, as well as global trade patterns, to demonstrate the complexity of the challenges that need to be overcome to deal with this problem more effectively.

Chapter 2 further explores these challenges by examining the global scope and impact of illegal fishing. The chapter begins with a discussion of what constitutes illegal fishing, followed by an overview of the geographical scope of the problem, species affected, and the characteristics of likely offenders. The chapter also provides a discussion on the ecological as well as economic and social impacts this crime has had and is likely to continue to have if nothing is done to stop it.

Chapter 3 looks at the conditions that facilitate illegal fishing. The chapter provides a review of methods used by illegal fishers to engage in these activities, such as using flags of convenience, ports of convenience, and transshipment, to circumvent fisheries regulations. The chapter also examines how illegal fishers take advantage of corruption, cultural tolerance of "rule bending," lack of political will to enforce fisheries regulations, and countries' inability to effectively monitor, control, and surveil their waters, to carry out their illegal activities.

Chapter 4 sheds light on why illegal fishing is happening, by examining it through the lens of environmental criminology. Drawing on the rational choice perspective and the techniques of situational crime prevention, the chapter guides the readers to think about illegal fishing as a crime of opportunity and as one that can be stopped if the opportunities for committing it are significantly reduced.

Chapter 5 tests the theoretical arguments made in Chapter 4 by analyzing data from 53 fishing countries, to reveal global patterns and clusters and the spatial dimensions of the problem, as well as to identify points of intervention.

Chapter 6 provides an overview of important international regulations, as well as innovative and promising regional and national interventions, that have been implemented to date to deal with illegal fishing. It highlights how these strategies fit within the techniques of situational crime prevention, albeit not categorized as such.

Lastly, Chapter 7 reemphasizes the importance of looking at illegal fishing as a crime and not as a fisheries management problem. It also argues that, to be able to stop illegal fishing, one must rely on data-driven, rigorous analyses to understand the micro-level factors and opportunity structures that lead to this criminal event. It also discusses a set of analytical strategies, conventionally used by crime analysts, to stop this global crime.

Chapter 1

Fishing and the Fishing Industry

A BRIEF HISTORY

Fishing in the Early Times

Humans have been fishing for millennia, and fishing is one of the oldest human activities. Fishing has played an important role in human livelihood since the prehistoric times, developing gradually when they moved from the unplanned collection of food to the more systematic way of harvesting and storing sustenance.[1] For example, during the Mesolithic Age (between 10,000 and 6,000 BC), forest areas expanded due to the warming of the climate, and this made hunting for bison, horses, and other land animals more difficult. Humans, consequently, began cultivating plants and breeding domestic animals. Importantly, they started settling permanently along the banks of rivers and lakes so that they could fish.[2] Many of the species that we catch today, such as Pacific salmon, have been harvested as early as 7,500 to 10,000 BC.[3]

Early humans caught fish by hand and with the help of weirs in rivers.[4] Later, they developed fishing gear to help them catch fish more easily and in larger quantities. In fact, fishing gear is among the oldest tools, and humans began using these tools earlier than any other artifacts surviving today.[5] Archeologists recovered the earliest fishing gear, the bone harpoon, from a site in present-day Congo, dating back to 90,000 years.[6] Modern humans are credited for having created successively cleverer artifacts thereafter, with examples including stone fish traps (75,000 years),[7] knotted fishing nets (35,000 years),[8] and fishhooks that were made of shells and neatly crafted stone, bone, and horn (23,000–18,000 years).[9]

In the Americas, a culture of fishers can be traced back to 4,000 BC, where early settlers lived by hunting, fishing, and collecting their food. A great majority of Indian tribes lived almost exclusively on fish, such as salmon, herring, cod, and halibut, because the upstream spawning migration patterns of the fish, as well as the concentrations of fish populations near the coasts, made it easier to harvest them.[10]

Early humans used marine species for other purposes as well. Cave paintings and rock carvings discovered in South Africa and Namibia, for example, suggest that humans made clothing, tents, and covers for kayaks, using sealskins. They also extracted oil from whales to burn in lamps; used fish scales to make sharp tools, such as knives and chisels; and made harpoons from animal fiber and leather straps.[11]

Ancient times, which saw a growth in education, art, culture, and the formation of governments, courts, and bureaucracies, also witnessed growth in human ingenuity in fishing. Ancient Egyptians, for example, already had extensive ichthyological and biological knowledge about more than 30 fish species, and this knowledge allowed them to not only catch fish in the wild but also cultivate the species in ponds. They also processed fish by drying, fermenting, and smoking.

The evolution of fishing methods, gear, and technology, such as iron fishhooks in Egypt[12] or the bronze-studded logs and net traps used by Mediterranean fishers to catch tuna,[13] gradually transitioned fishing from a mere sustenance activity into an economic undertaking. According to the writings of Greek and Roman historians Herodotus and Aelianus, fishing was so important and prevalent in Ancient Egypt that there was a set system of taxation on the catches and fees for fishing rights in place. Fish could also be used as payment in return for labor.[14]

Archeological evidence suggests that fishing was an established commercial activity in the Mediterranean Sea, Northern Black Sea, and Atlantic coast (e.g., Spain, Morocco, and Portugal) as far back as the fifth and third centuries BC. Early Mediterranean civilizations already had sophisticated methods, such as elaborate traps known as "tonnara,"[15] to capture bluefin tuna—a fish that has become an economically significant commodity since the earliest recorded history in the region.[16] Fish stocks were so important that these may have been one of the primary reasons (the most important being wheat) why the Greeks colonized the Black Sea region as early as the seventh century BC.[17]

Roman coastal towns housed factories that can be traced to today's Italy, Spain, North Africa, and Turkey.[18] At these factories, they fermented and dried sardines and anchovies and made these into *garum*[19]—a pickled fish paste used as salt or soy sauce on meat, fish, and vegetables and in many other dishes in Roman cuisine.[20] High-quality *garum* made from small tuna was exported in special jars throughout the vast Roman Empire, with a "signature" spine of tuna placed in the bottom of these jars to attest to the quality of the sauce. On the coast of the Bay of Naples, *garum* manufacturing was a principal industry. Interestingly, the first recorded decline due to overfishing is also credited to Romans during the first century AD, who eventually had to travel farther distances to obtain fish for the *garum* industry after having depleted their own resources.[21]

Ancient Chinese were quite distinctive in their fishing methods. For example, they fished with tamed and trained cormorants. The fishers released these birds from their boats and rafts to catch the fish. But these birds had rings placed around their throats to prevent them from swallowing the larger pieces of their prey, thus delivering the fish to their owners.[22] Ancient Chinese engaged primarily in freshwater fishing; however, frequent typhoons and piracy in the fishing regions forced them to develop and later rely extensively on aquaculture to meet their daily protein needs.[23]

Fishing in the Middle Ages

In the Middle Ages, herring, cod, and whale were among the most important seafood for consumption and trade. Between the fifth and fifteenth centuries, the herring trade flourished along the coasts and navigable rivers in Northern Europe.[24] The English, French, and Flemings actively fished herring in the North Sea and the English Channel, while the Danes and Norsemen fished around the British Isles.[25] The Hanseatic League[26] already had a monopoly over the Baltic herring trade by the fifteenth century.[27] The herring fishery became the first fishery to reach the level of a real industry.[28] However, as early as the fourteenth century, North Sea herring was already considered to be heavily overfished.[29] In the sixteenth century, the Scanian herring fishery, which played an important role in the daily lives and economy of the people in Sweden and Denmark, collapsed due to overfishing.[30]

Cod was among the most commonly available fish in Europe during the Middle Ages. The Basques applied the salting techniques that they had already perfected with whale meat to cod. This gave them a great market advantage because with the salted cod they were able to travel longer distances, expanding into more remote markets.[31] English fishing vessels carried out cod fishing along the shores of the English Channel. Fishing for cod was later expanded along the coasts of the North Sea.[32]

With the spread of Christianity throughout Europe, cod became "almost a religious icon, a mythological crusader for Christian observance."[33] Catholic dietary laws imposed abstinence from eating red meat (or the flesh of warm-blooded animals) about 150 days a year, which included every Friday (because Jesus reportedly died on a Friday), the 40 days of Lent before Easter, and shorter periods before other important religious holidays.[34] These religious ordinances created an increased demand for fish (because they were cold-blooded), including herring and cod, and this increased demand promoted fish trade in Europe.[35]

The most prominent cod fishery, however, developed in the Northwest Atlantic of Newfoundland. Basque sailors are credited to have reached these waters in the early fourteenth century in search for whales, which had become scarce in the Gulf of Biscay.[36] Increased tensions over Icelandic cod fisheries, such as those between England and the Germans of the Hanseatic League,[37] and later cod shortage in Europe set into motion the European exploitation of the New World fisheries soon thereafter.[38] The French, British, and Portuguese set sail in search for new cod resources. This new exploration led to the discovery of huge cod populations by John Cabot in the fifteenth century, who marveled that the cod were so plentiful that they virtually "blocked his ship." This also led to the eventual European settlements at the Grand Banks of Newfoundland.[39] Here, fishers mainly fished in waters close to shore, using small craft and traditional techniques.

The Basques have been catching whales along the northern Atlantic coast of Spain since the seventh century.[40] Initially, they gathered whales that were stranded along the coasts. Later, with the construction of small boats, they were able to sail farther distances and use harpoons to catch them.[41] The advances in sailing and navigation allowed Europeans to whale farther north, reaching the high Arctic around Spitsbergen by the seventeenth century.[42] However, by 1690, whale stocks in the eastern Arctic were depleted. Whales were used for their meat, with the whale tongue being among the most prized part of the mammal.[43] Whale blubber was processed into oil to be used in oil lamps. Its ribs were used as construction material for huts, while its other bones were used to make utensils, tools, and other artifacts, such as stiffening material in corsets. Whaling was also a common and important economic activity along the coasts of New England, where early Yankees whaled much like the Basques in Europe.[44]

Attempts to regulate fisheries date back to the seventh century. These regulations contained restrictions of fishing in time and place, the gear used, and species to be caught. These regulations also imposed strict controls over selling of fish and were in response to the growing concern for the resource destruction and overfishing that was a result of Europe's growing populations and consequent increased demand for fish.[45] However, it was not until the thirteenth century when the first large-scale fisheries ordinance was issued (by French King Philip IV in 1289), and it reflects vividly the impact overfishing had on the resources in the region:

> today each and every river and waterside of our realm, large and small, yields nothing due to the evil of the fishers and the devices of [their] contriving, and because the fish are prevented by them from growing to their proper condition, nor have the fish any value when caught by them, nor are they any good for human consumption, but rather bad, and further it happens that they are much more costly than they used to be, which results in no moderate loss to the rich and poor of our realm.[46]

Depleting freshwater fisheries, however, did not impact the demand for fish. More sophisticated vessels were soon developed during the Middle Ages, and this allowed fishers to not only undertake longer voyages but also catch dramatically more fish.

The Two Periods of Industrialization and Expansion of Fisheries Worldwide

Today, fishers employ a wide variety of sophisticated techniques and vessels and fish in expansive fishing grounds. This is primarily due to the human population growth, which necessitated the shift from catching small quantities of fish to developing means to catch fish in large amounts.[47] The growth in mass harvesting of fish, however, would not have been possible if not for the technological and scientific developments that took place during the Industrial Revolution.

The Industrial Revolution, which began in England and spread first to Western Europe and then to the United States, occurred between the late eighteenth and early twentieth centuries. This period is marked by a rapid growth in mechanization and industrial production. The two stages of the Industrial Revolution (1770–1870 and 1870–1914) accelerated the overall societal growth. People moved from working primarily in agriculture to being employed in factories. During this period, the inventions of new technologies, the development of steam and later petrol engines, and the rise of steam trains and other steam-powered machines led to a remarkable increase in mass production of goods and the overall efficiency in the distribution of these goods. The Industrial Revolution also marked the beginning of an unprecedented expansion of fisheries worldwide and the growth of fishing into a multibillion-dollar industry. Two distinct periods mark this growth.

The First Industrialization Period (1870–1950)

No century has seen as remarkable an increase in the number of fishing boats and gear as the last decades of the nineteenth century. This growth has been especially noticeable in Europe and North America.[48] James Watt had introduced the first serviceable steam engine in 1769, and about 100 years later these machines were applied to the propulsion of fishing vessels in Europe. The introduction of steam drifter vessels and, a short time later, motors instead of sails, made vessels completely independent of wind conditions. Initially, the steamers were used as tenders for transporting the fish from the fishing fleets to the ports. They were also used as tugs for towing the sailing ships to and from the fishing grounds either when the wind was blowing from unfavorable directions or when it was lacking throughout.[49]

During this period, a key invention that led to significant increases in fishing power was the steam-powered trawler,[50] which spread rapidly across the North Sea in the 1880s and around the world by the early 1900s (from New Zealand to South Africa to Chile). This development allowed for massive catches of herring, flatfish, and cod.[51] This increased mobility and catching efficiency of vessels also allowed the harvesting and transport of fresh fish over large distances. The eventual increased mechanization led to significant expansions of fishing activities farther into offshore fishing grounds and in search for larger fishes, broadening the opportunities for harvesting and landing a steady supply of fish to markets.[52] In fact, the technological expansion into the fisheries sector was driven by the increasing opportunities to profit from this growing industry,[53] as colonial merchants and capitalists saw fish as an infinite source of revenue.[54]

This period also marked the beginning of the marketing of fresh fish due to the availability of cold storage.[55] The use of ice as a means of preserving fish (instead of drying and salting) became more common, with the first reported use dating back to the 1790s[56] when fresh salmon was transported from Scotland to London.[57] By the mid-1800s, the British used ice cutters to cut ice from rivers and lakes in the winter, storing them in expansive storing houses along the shores.[58]

Around the end of the nineteenth century, world fish catch was approximately 8–10 million tonnes per year. The most important fisheries were in Russia and Japan, with annual catches of one million tonnes each, followed by China and countries in Southeast Asia. Japan's fisheries expansion began in the 1890s, with the Sino-Japanese War (1894–1895) and Russo-Japanese War (1904–1905) marking the first phase of fleet motorization in the country.[59] Japanese fleets actively fished in waters around Taiwan, the Tonkin Gulf of Vietnam, and the South China Sea.[60] China was not as successful as Japan in expanding its fishing operations in Asia until the Chinese Communist Party came to power in 1949. This was primarily attributed to the fall of the Ch'ing dynasty (in 1912), the civil war, and the subsequent Japanese invasion.[61]

Around the 1920s and 1930s, the world fish catch reached approximately 20 million tonnes. In 1936, fish landings were 17 million tonnes, with the most important landings being in Japan, the United States, Korea, the USSR, Norway, and Great Britain.[62] Europe contributed approximately 500,000 tonnes to the global catch from the cod fisheries alone. However, this dramatic expansion of fishing, and especially of trawling, led to the significant decline of "prime" species groups, such as flatfishes. This was primarily due to the heavy exploitation of fisheries before World War I.[63] The war stalled fishing activities in much of Europe for a short period of time, but its aftermath led to increased fish catches resulting in the depletion of several fish stocks in the North Atlantic shortly thereafter. At the end of the 1920s, the fisheries and the fishing industry suffered significantly from the world economic crisis.[64]

The Second Industrialization Period (1950–1980)

The second industrialization period, marked by yet another expansion in fisheries catches worldwide, corresponds to the aftermath of World War II, and this growth was driven by the postwar economic recovery.[65] Postwar prosperity, improved fishing fleets, and the development of new fishing technologies (such as fish-finding sonar and lighted synthetic nets) improved processing and preservation technologies onboard fishing vessels (such as onboard canning and flash-freezing), coupled with the rapidly growing world markets for wild fish, collectively contributed to this dramatic expansion.[66] Fisheries industrialization in developing countries[67] brought about by the UN Food and Agriculture Organization (FAO) projects that included technology transfers and the establishment of bilateral development aid also contributed to this growth.[68]

The second industrialization period is considered the most important phase of the expansion of fisheries worldwide. This period is marked by considerable improvements in fishing techniques and fishing gear, as well as the growth in the size of the vessels and gear.[69] At the time, the USSR had the largest number of factory trawlers capable of traveling great distances and fishing at great depths,[70] but trawlers were also common and widely used in Europe and North America.[71]

The aftermath of World War II also led to the rapid growth of the Japanese fishing industry, and this was primarily due to the need to fill the gap of food deficit that emerged in the late 1940s.[72] China's first fisheries expansion period (1950–1959), also known as "the period of initial development," coincided with the beginning of this industrialization period.[73] Other countries, such as South Korea, India, Philippines, Malaysia, Indonesia, and Thailand, experienced similar expansions of their fisheries, and this was also due to the availability of modernized fishing technologies and gear, as well as motorized fishing vessels.[74] It was not until after World War II that the fishing capacity in Australia and New Zealand was notably expanded.[75] Neither was this expansion significant in the South American–Caribbean region until the same period.

While the coastal African countries depended heavily on fish for livelihood for centuries, the development of the African fishing industry was slow, and before the 1950s, local fishers primarily used small artisanal vessels to exploit the fisheries in the continent.[76] Since the late 1950s, semi-industrial and industrial fisheries developed parallel to the artisanal fisheries.[77] By the late 1960s, Western and Southern subregions in Africa, which accounted for more than 80 percent of the continent's marine resources,[78] gradually expanded the exploitation of their natural resources.[79]

Overfishing became a problem at the global scale and led to the depletion of many of the world's fish stocks as early as the 1960s.[80] The worldwide expansion of fisheries during this period led to the collapse of some important fish stocks, including the Californian sardine, North Atlantic herring, Peruvian anchoveta, North Sea mackerel, and Atlantic menhaden.[81] Using heavily mechanized purse seines to capture herring in the North Sea, for example, led to the overfishing of the species to the point of its biomass collapse by the 1960s.[82] However, quite often, "environmental effects" and not overfishing were suggested as the cause of these collapses, an example of which is the El Niño effect that was implicated as the main cause for the collapse of the Peruvian anchoveta. This explanation allowed commercial fishers to continue with business as usual, without the fear of significant repercussions.[83]

The problem of overfishing, coupled with the growing sense of failure of the international community to properly manage marine resources, was a primary reason for the establishment of the unilateral declaration of the exclusive economic zones (EEZs) by many countries in 1974 at the Third United Nations Conference on the Law of the Sea (UNCLOS) in Caracas, Venezuela. These EEZs extended state jurisdictions over marine resources to 200 nautical miles from the coast. This change in access forced many countries that had been operating distant water fleets in other countries' waters to redirect their operations to their own EEZs and to international waters.[84]

The UNCLOS attempt to address the problem of illegal fishing in the 1970s was not very successful, as many commercial fisheries only continued to be exploited beyond their yield capacity. For example, in Canada, within less than 10 years, Atlantic cod populations in Newfoundland declined to the point that they could yield only 300,000 tonnes of cod by 1975, a decline of 500,000 tonnes from prime years.[85] While this was a sign of intense fishing pressure and significant overfishing, the overharvesting of this important commercial species did not stop. Instead, investments poured into constructing factory trawlers (or draggers) that could fish farther offshore. These draggers not only destroyed critical habitat but they also targeted spawning cod stocks, eventually leading to catastrophically low numbers of cod for the first time in 400 years.[86] This left the Canadian Minister of Fisheries with no choice but to pass the 1992 Northern Cod moratorium banning the harvesting of cod in Newfoundland waters, the Grand Banks, and most of the Gulf of St. Lawrence.[87] Based on 1994 estimates, a meager 1,700 tonnes of cod biomass remained in the fishery[88] that had for centuries been feeding and providing employment to fishers and coastal communities. This ban led to the loss of 40,000 jobs, striking a significant economic blow to the coastal communities in the region that had been relying heavily on this marine resource for centuries.[89]

The severe depletion of whales over time, which started from the seventeenth century and continued until the twentieth century, coupled with the concerns that markets were being flooded with whale oil, led to the establishment of the International Whaling Commission (IWC) in 1946.[90] The initial efforts to regulate the catch were deemed unsuccessful, as individual countries disregarded these regulations and took advantage of the lack of enforcement of these regulations. In 1985, the IWC imposed a moratorium with a zero quota on all commercial whaling. However, a limited number of countries, including Norway, Iceland, and, notably, Japan, still hunt whale.[91] In 1986, Japan launched a "scientific whaling" program, which was a cover for its commercial whaling that continues today. Norway complied by the IWC's moratorium requirement up until 1993, when it launched minke whale harvesting operations within its waters that allowed its whalers to fish for about 1,000 whales a day since then. Iceland engaged in "scientific" whaling up until 1992, withdrew from the IWC from 1992 to 2004, and rejoined the IWC in 2004 only with an objection to the whaling moratorium, which technically allowed them to hunt whale.

Fisheries scholars recognize three different periods of the human impact on the marine fisheries: aboriginal, colonial, and global. Aboriginal exploitation refers to the use of nearshore coastal resources for subsistence use, which was made possible with the use of more rudimentary gear, watercraft, and extractive technologies. Colonial exploitation involved the systematic use of distant water resources, which developed as a result of the systematic depletion of resources in the coastal and shelf seas of the European countries. The colonial exploitation also marked the beginning of the development of fishing into a market economy. Global use referred to the geographically more expansive and intense exploitation of fisheries resources, characterized by the patterns of extensive consumption and frequent exhaustion (and consequent substitution) of these resources.[92]

FISHERS AND THE FISHING INDUSTRY: A SHORT SURVEY

Fishing as an Economic Activity

The seafood industry comprises three distinct sectors that engage in catching, processing, and marketing of fish and fish products. The catching sector includes fisheries and fish farms. Fisheries harvest fish from the wild, while fish farms cultivate fish under controlled condition. Figure 1.1 shows a structural diagram of the seafood industry.

Commercial fisheries can be either large-scale, or industrial, or small-scale, or artisanal. Large-scale commercial fisheries where species are caught for reduction purposes, such as for fishmeal or manufacture of fish oil, are generally referred to as reduction fisheries, and these are the largest fisheries in the world. Reduction fisheries almost exclusively target small species inhabiting the upper layer of the sea, also known as pelagic fishes. Such industrial species as sprat, capelin, and horse mackerel are used in fish oil that is, in turn, used in a range of products that include margarine and biscuits.[93] The Peruvian purse seine fishery for anchoveta, which is considered the world's largest fishery, is an example

Figure 1.1 The Structure of the Commercial Sector

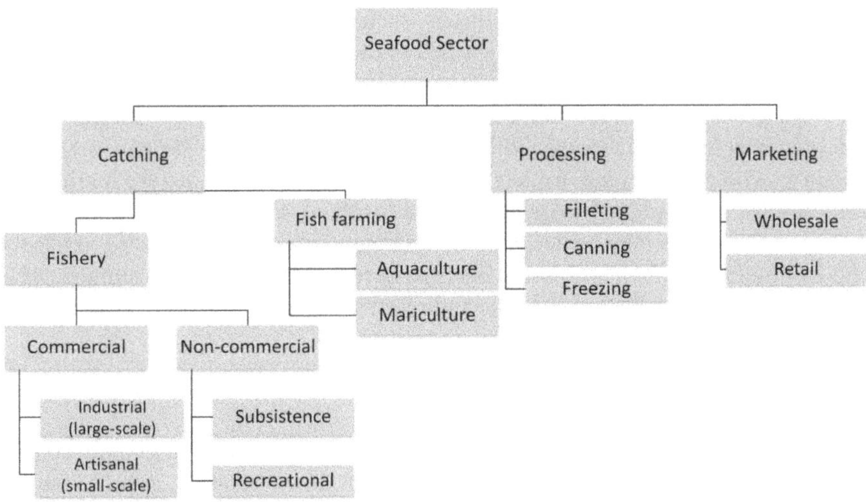

of a reduction fishery. Other large-scale commercial fisheries catch species that are sold in markets for direct human consumption.

Artisanal fisheries are small-scale commercial fisheries, where caught fish are generally sold in local markets or sometimes used for exportation. Vessels used in artisanal fisheries range from small one-person canoes in poor developing countries to trawlers, seiners, or long-liners that are over 20 meters (approximately 65 feet) long, which are more common in developed countries.[94]

Fish can also be caught for noncommercial purposes, such as in the case of subsistence and recreational fisheries. Subsistence fisheries are usually small scale, and the fish caught here are generally shared or consumed directly by the families of fishers.[95] These fisheries are those found close to shore. Fishers generally use small fishing vessels to catch fish from these types of fisheries. Fish caught by subsistence fishers are not sold. Local fishers may sometimes exchange fish for other goods or services.[96] In many parts of the world, such as the Western Pacific, subsistence fisheries constitute approximately 80 percent of inshore fisheries catch.[97]

The recreational sector does not include sale or trade of fish. Instead, it includes licensing for recreational fishing or sport fishing, that is, fishing for competition or for pleasure. This sector also consists of enterprises that manufacture and retail fishing equipment, apparel, books, and magazines, as well as engages in the design and sale of recreational fishing boats. The sector contributes significantly to the economies of developing countries. For example, in the United States, approximately 60 million anglers spend approximately $46 billion annually on fishing equipment, transportation, lodging, and other expenses associated with their sport. The industry itself generates approximately $115 billion in profit in the United States and supports more than 828,000 jobs.[98]

Fish farming includes aquaculture and mariculture. Aquaculture, also known as aquafarming, refers to the cultivation of marine and freshwater species for consumption under controlled conditions, such as in ponds or in tanks. Mariculture involves the raising of marine species in their natural habitats, such as freshwater or saltwater.

Fish processing, which may range from handling raw fish to manufacturing fish products, involves freezing, filleting, or canning fresh fish for onward distribution to wholesale or retail outlets. The marketing chain includes the marketplaces used for the trade and sale of seafood products either wholesale or retail. These marketplaces may be designed to facilitate this trade either between fishers and fish merchants on a wholesale basis or between fishers and consumers, who can be either buyers for food chains and retail stores or individual consumers.

Global Fisheries Catches and Trade at a Glance

Fish and fish products are a valuable source of protein and play an integral role in the world population's intake of animal protein. Fish comprise approximately

20 percent of intake of animal protein for approximately three billion people worldwide.[99] Both fisheries and aquaculture play an important role.

The global fisheries catches and aquaculture production in 2014 were about 167 million tonnes of fish with an estimated value of about $308 billion.[100] Fisheries make around 56 percent of the overall fish production. Of these, 82 percent come from marine fisheries, and the remaining catches originate from inland fisheries. Aquaculture makes up the remaining 44 percent of the total production, with 47 percent coming from inland waters.[101]

Asia is the largest consumer of fish, accounting for approximately 99 million tonnes or two-thirds of total world consumption, which translates into 23 kg per capita consumption. The highest per capita fish consumption in the world is in developed countries, accounting for an average of 36.8 kg per capita, whereas the fish consumption in low-income food-deficit countries was about 7.6 kg per capita in 2013.[102]

The principal global fisheries have largely remained the same, except for the Peruvian anchoveta fishery, which, due to the significant drop in anchoveta catches in 2014 (since 1998), is no longer one of the top five fisheries in the world. In 2014, the top producers were China, Indonesia, the United States, India, and the Russian Federation. Of the total of 93.4 million tonnes caught in 2014, about 37.4 million tonnes or about 40 percent were from the fisheries of these top five countries alone. When the top 10 global producers are considered (adding Myanmar, Japan, Peru, Vietnam, and the Philippines), fisheries yield reaches to 54.1 million tonnes or about 58 percent of combined world fisheries yield.[103]

In 2014, developing countries accounted for approximately 54 percent in value and about 60 percent of overall fisheries exports by quantity (combining fisheries catches and aquaculture production). In fact, the exports from developing countries of fish and fish products for human consumption almost doubled from 32 percent in 1980 to 54 percent in 2014. The top 10 exporters, which include both developing and developed countries, accounted for 53 percent of the world's exports in value, with China, Norway, Thailand, Vietnam, and the United States making the top five list.[104]

Table 1.1 shows the top exporting and importing countries of fish and fish products in 2014. The top 10 exporters accounted for about 53 percent of all the exports in the world, and the top 11 importers accounted for 59 percent of the global imports.[105] China was the top exporter, exporting twice as much as the second largest exporter, Norway. The top five exporters accounted for 34 percent of the total exports in the world. The United States, the top importer in 2014, imported 14 percent of the world's total fish and fish products. The top five importers accounted for 41 percent of the world's total imports.

Fisheries target over 1,600 species; however, a total of 23 species comprised about 40 percent of the total capture in 2014. Table 1.2 lists the top 10 principal species from fisheries in order of importance. The total catch for these 10 species was about 21 million tonnes in 2014, or approximately 26 percent of the world's total catches.[106]

Table 1.1 Top Importers and Exporters of Fish and Fish Products in 2014

Exporter	USD Billion	Importer	USD Billion
China	21.0	United States	20.3
Norway	10.8	Japan	14.8
Vietnam	8.0	China	8.5
Thailand	6.6	Spain	7.1
United States	6.1	France	6.7
Chile	5.9	Germany	6.2
India	5.6	Italy	6.2
Denmark	4.8	Sweden	4.8
Netherlands	4.6	United Kingdom	4.6
Canada		Republic of Korea	4.3
Top 10 subtotal	**77.8**	**Top 10 subtotal**	**83.45**
Rest of world total	**70.35**	**Rest of world total**	**57.17**
World Total	**148.15**	**World Total**	**140.62**

Table 1.2 Top 10 Principal Species from Fisheries around the World in 2014

Common Name	Scientific Name	Catch (million tonnes)
Alaska pollock	*Theragra chalcogramma*	3.2
Anchoveta	*Engraulis ringens*	3.1
Skipjack tuna	*Katsuwonus pelamis*	3.1
Sardinellas nei	*Sardinella* spp.	2.3
Chub mackerel	*Scomber japonicus*	1.8
Atlantic herring	*Clupea harengus*	1.6
Yellowfin tuna	*Thunnus albacares*	1.5
Scads	*Decapterus* spp.	1.5
Atlantic mackerel	*Scomber scombrus*	1.4
Japanese anchovy	*Engraulis japonicus*	1.4

The world aquaculture production has seen a steady increase, with an annual growth rate of 5.8 percent. While in 1980, the global aquaculture production was about 4.7 million tonnes, this number was 73.8 million tonnes in 2014, reaching more than a 1,000 percent increase in a little over 30 years.[107] The top 10 producing countries accounted for approximately 89 percent of this production, with Asia accounting for 89 percent by value of the world's farmed fish for food. The major producers in Asia are China (producing 60 percent of the world's production), India, Vietnam, Indonesia, Bangladesh, Thailand, Myanmar, the Philippines, and Japan.[108]

World aquaculture production targets over 100 species from different groups that include fish, crustaceans, and mollusks, as well as aquatic plants and other aquatic organisms. In 2014, finfishes accounted for approximately 68 percent of total production, followed by mollusks (22 percent), crustaceans (9 percent), and other aquatic species (1 percent), such as softshell turtles, sea cucumbers, frogs, jellyfishes, sea squirts, and sea urchins.[109]

Fishers and Their Workforce

A number of actors make up the fishing industry. Among these, the most important are the fishers, that is, people who are engaged in the catching of fish. Fishers may or may not own the vessel they operate. Oftentimes, fishers are employees onboard the fishing vessels, and they work for the vessel owner, also known as the operator. The vessel "owner" is generally the person or entity that has a direct control over the vessel's operations. The true beneficial owners of the vessel may sometimes be anonymous, but they are often closely linked to the person overseeing the fishing operations.[110]

As of 2014, an estimated 57 million fishers and fish farmers worked in the fishing industry worldwide, of which 84 percent worked in Asia, 10 percent worked in Africa, and 4 percent worked in Latin America and the Caribbean.[111] The top five countries that collectively contribute to approximately 40 percent of the global workforce of fishers and fish farmers include China (14 million), Indonesia (6 million), Taiwan (330,000), Mexico (271,000), and Japan (203,000). A vast majority of these fishers and fish farmers work in artisanal fisheries, which contribute to almost 50 percent of the world's catch for direct human consumption.[112]

HOW ARE FISH CAUGHT?

An Overview of Fishing Gear

As seen from the review presented earlier, new technologies to catch fish and other marine species consistently developed to address the growing human need for food. This development significantly expanded following the introduction of the steam engine, which made it possible to not only travel farther distances and remain in the sea for extensive periods but also catch more fish. Today, the fishing gear is equipped with sophisticated technologies, such as underwater cameras, that provide a huge advantage to fishers.[113]

There are a total of about 81 fishing gear classified under 11 different categories as per the International Standard Statistical Classification of Fishing Gear. These categories can further be classified as either mobile/active or static/passive, and this depends on whether they are fixed to the seabed by boats or dragged along.[114]

Each type of fishing gear is designed to capture certain types of fish species, and based on this, they are grouped into three major groups: those that target groundfish, or demersal fishing gear; those that target large pelagic fish (i.e., fish living closer to the surface); and those that target small pelagic fish. Among the major gear used to capture groundfish are bottom trawls, gillnets, and hooks and lines, which are used to capture cod, haddock, flounders, and other species. Purse seines and long lines are used to capture large pelagic species, such as tuna and billfishes; and midwater trawls and purse seines are among the gear used to capture small pelagic fish, such as mackerels and pilchards.

An Overview of Fishing Techniques

Fishing techniques (of which there are about 80) are classified by the species they target. There are nine major such techniques, of which four are used to catch exclusively demersal species and three to catch only pelagic species. These include the following:

- American boat purse seining
- Bottom pair trawling
- Danish seining
- Drum seining
- European boat-operated purse seining
- Midwater pair trawling
- Pair seining
- Scottish seining
- Two-boat-operated purse seining

Among all the fishing techniques, trawling is considered the most important and one of the most efficient methods. It can be carried out from very shallow waters up to a depth of 2,000 meters.[115]

American boat purse seining and European boat-operated purse seining target pelagic species that travel in dense schools. Bottom pair trawling and Scottish seining are used to catch demersal or bottom species. Bottom pair trawling is used to catch European hake, Atlantic cod, flatfish, and shrimps and is operated up to 800 meter depths. Drum seining targets salmon and herring and is seasonal, depending on when fish aggregate into large, dense schools. Unlike these techniques, which target either pelagic or demersal species, midwater pair trawling is used to target both and is used to catch such known demersal species as Atlantic herring, European pilchard, European sprat, hake, and sea bass.[116]

Aquaculture techniques have seen a rapid development in the past several decades, and this is not only due to the need to expand the growing of aquatic animals and plants but also due to the developments in engineering and technology. Species are farmed in a variety of environments, and these range from small ponds and open net pens or cages to raceways and indoor hatcheries.[117] Ponds are fed by natural rainfall or ground water, or by diverting a stream, river, or an irrigation canal. Indoor rearing of species involves more controlled conditions, where the water quality is monitored and water condition is maintained so as to facilitate the reproduction and growth of some species. Net pens or cages are artificial structures that float in protected coastal or inland waters, such as fenced areas of the sea or in lakes. Raceways are artificial channels usually constructed of concrete. These channels are built near streams, reservoirs, or other water sources that allow for a constant flow of water and are constructed with an inlet and outlet to facilitate this flow and maintain water quality.[118]

An Overview of Fishing Vessels

Fishing vessels are boats or ships used to catch fish in the sea, lake, or river. There are about 4.6 million fishing vessels worldwide.[119] Of these, 41 percent are small vessels without engines, also known as nonmotorized. The remaining 59 percent are engine-powered vessels. Of the motorized fishing vessels, 75 percent are from Asia, with China, Indonesia, and India being the top three countries in terms of the size of their fishing fleets. The remaining vessels are from Latin America and the Caribbean (8 percent), Africa (7 percent), and Europe (4 percent).[120]

Approximately 85 percent of the motorized fishing vessels in the world are less than 12 meters in length, that is, small scale, and only about 2 percent of all motorized fishing vessels are 24 meters and larger, or industrialized fishing vessels. The larger vessels are predominantly from the Pacific region, Oceania, Europe, and North America.[121] The greatest majority of fishing vessels, or approximately 3.2 million, operate in marine waters, and the remaining vessels operate in inland waters.

Fishing vessels are classified into different categories based on their size and the techniques used onboard these vessels. For example, a trawler uses trawls or trawl nets to catch fish, while long-liners use fishing lines that have baited hooks that are attached to the main line by branch lines or "snoods." The UN FAO classifies all fishing vessels into 10 categories. Each of these categories also contains subcategories of vessels, which are grouped based on their functionality (see Table 1.3). The following are short descriptions of each vessel type, their fishing methods, and examples of the types of fish caught using these vessels.

Trawlers. These are vessels designed to tow cone-shaped trawl nets behind them (see Figure 1.2). These vessels actively drag or pull these nets at various depths that may range from just below the surface of the water to just off the seafloor.[122] There are various types of trawlers, which range in size and functionality. For example, pair trawlers are about five meters in length and can handle approximately five gross in register tonnes of fish. Meanwhile, freezer trawlers are 40 meters long and can handle about 370 gross in register tonnes of fish.[123] Trawlers target such species as cod, flounder, and shrimp.

Table 1.3 Fishing Vessels by Type (as per UN Classification)

Trawlers	**Long-Liners**	**Other Fishing Vessels**
Factory trawlers	Freezer long-liners	Lift netters
Freezer trawlers	Factory long-liners	Lift netters using boat-operated net
Wet-fish trawlers	Wet-fish long-liners	Other lift netters
Outrigger trawlers	Other long-liners	Vessels using pump for fishing
Beam trawlers	**Other Liners**	Platforms for mollusk culture
Other trawlers	Jigging line vessels	Recreational fishing vessels
Purse Seiners	Handliners	**Trap Setters**
Tuna purse seiners	Pole and line vessels	Pot vessels
Other purse seiners	Trollers	Other trap setters
Other Seiners	Other liners	**Multipurpose Vessels**
Sein netters	**Gillnetters**	Trawlers-purse seiners
Other seiners	**Dredgers**	Other multipurpose vessels

Figure 1.2 Trawler (Roza Minasyan)

Figure 1.3 Long-liner (Roza Minasyan)

Long-liners. These are vessels that have central fishing lines or mainlines attached to them (see Figure 1.3). These lines are equipped with baited hooks that can range from several hundred to several thousand.[124] Depending on where these lines are placed in the water, long-liners can be classified into pelagic long-liners and demersal long-liners. Pelagic long-liners set the lines near the surface of the water and target open ocean fish, such as tuna and swordfish. Demersal or bottom long-liners float their hooks along the seafloor and target species that live on or near the bottom, such as cod and halibut. Long-liners can also use traps, instead of hooks, attached to the lines, which are used to catch crabs in deep waters.[125]

Purse seiners. These are vessels that have seines or dragnets attached to them (see Figure 1.4). These seines hang vertically in the water, with the bottom edge of the net held down by weights and its top edge buoyed by floats. Fish are encircled into the center of the net, the bottom of which is then pulled to close like a drawstring purse. Purse seiners target sardines, squids, and tuna used for canning and sashiming.[126]

Gillnetters. These are vessels that use curtains of netting suspended by floats and weights (see Figure 1.5). The nets are invisible to fish, which make it easy for

Figure 1.4 Purse Seiner (Roza Minasyan)

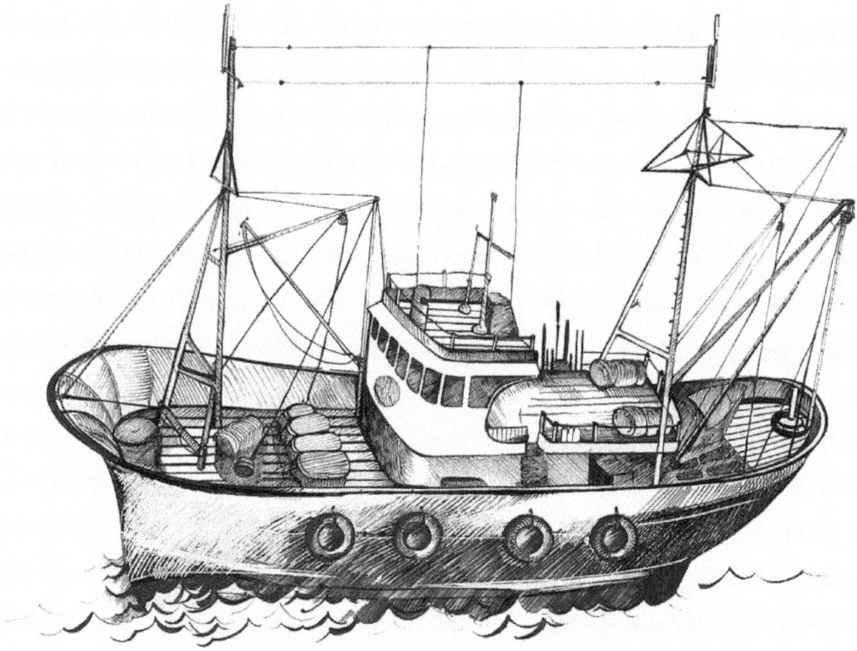

Figure 1.5 Gillnetter (Roza Minasyan)

Figure 1.6 Dredger (Roza Minasyan)

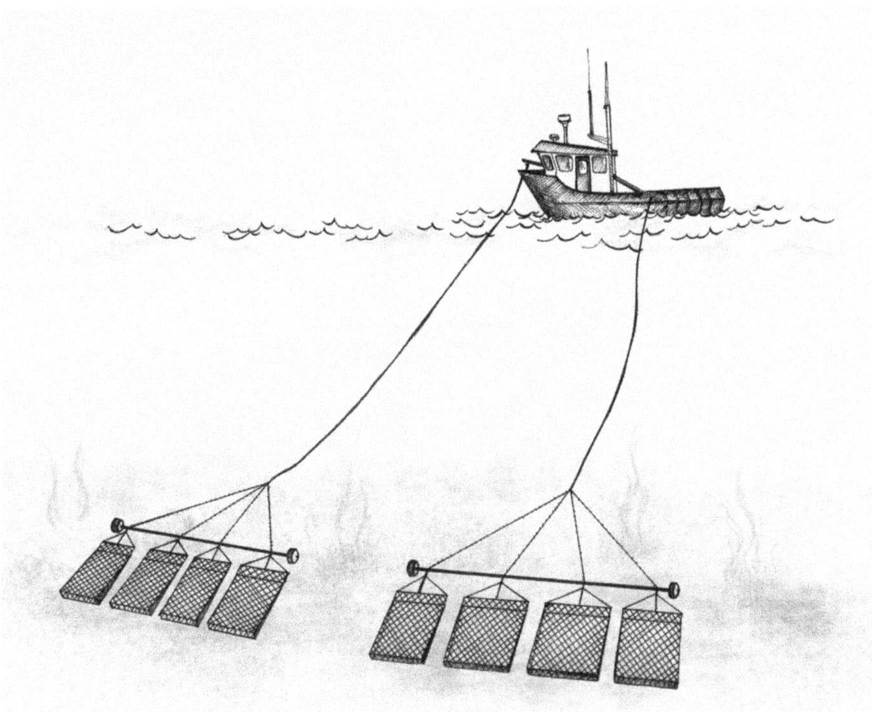

them to get trapped when swimming into these nets. Fish are tangled, held by the mesh around the body, or trapped by mesh slipping behind the opercula when they swim into these suspended nets.[127] Gillnetters target such species as sardines, salmon, and cod.

Dredgers. These are vessels that are used to drag dredges over the seabed (see Figure 1.6). These dredges are used to dig shellfish and other mollusks from the ground.[128]

Trap setters. These are vessels that have traps or pots attached to them, which are wire or wood cages that are submerged into the water (see Figure 1.7).[129] These traps or pots are used to catch bottom-dwelling fish, such as sablefish, as well as lobsters, crabs, and shrimp.

CONCLUDING REMARKS

The brief overview provided in this chapter showed that the role of opportunity in building fishing into a large-scale industry should not be ignored. The development of fishing from a subsistence activity into a multibillion-dollar industry was

Figure 1.7 Trap Setter (Roza Minasyan)

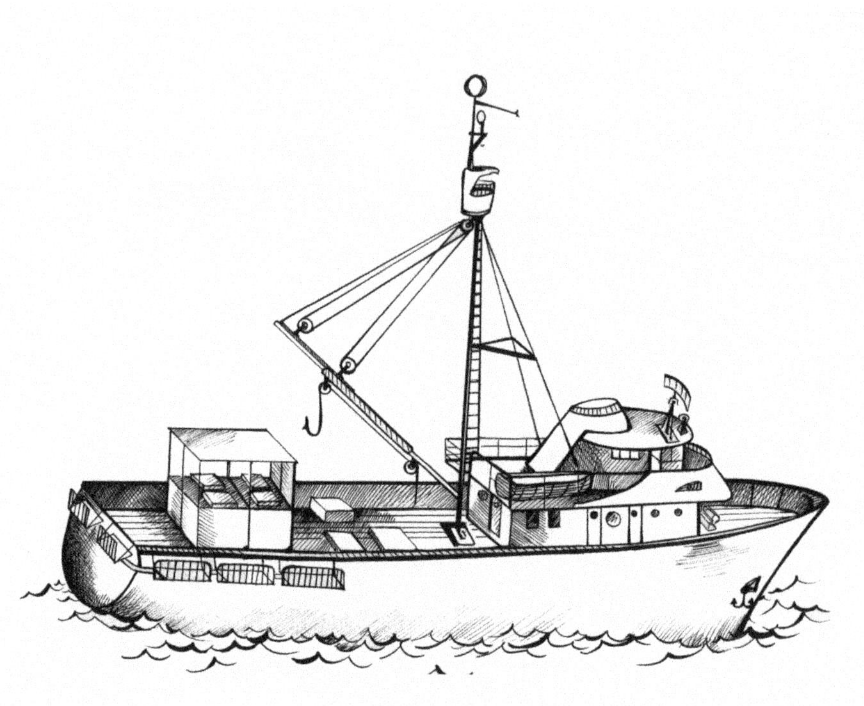

made possible because of the relentless advancement of fishing gear that allowed fishers to exploit new fishing grounds and catch new species. When the opportunities in certain fisheries or for harvesting particular species diminished due to overfishing, shifts in search for new opportunities followed. Humans exploited freshwater fish at first, but the depletion of resources in these waters only led to the search for opportunities in seas and more expansive waters. Deepwater fish, such as Patagonian toothfish and large tunas,[130] became an important commodity in the 1960s[131] after the newly developed technologies allowed fishing in deeper waters. The evolution of means of removal, preservation, and transportation of fish essentially contributed to the global fisheries expansion into an essentially worldwide market economy. The insatiable desire to capitalize on this commodity, coupled with the erroneous belief in its limitless supply, however, had tragic consequences, exemplified by the 1992 collapse of the centuries-old cod fisheries off the coast of Newfoundland. However disturbing, the pattern of human overexploitation of fisheries nevertheless continues today.

The discussion on the global trade trends provided in this chapter illustrated the fishing industry's significant role in today's food markets worldwide. The complexity of the industry, with its different sectors and actors involved in each sector; gear used; and methods employed to target specific types of fish species speak to the challenges and the enormity of the task of properly regulating such an industry so as to not only feed the billions of people who depend on fish as a source of protein but also to ensure that these fisheries are sustainably caught to feed billions more in centuries to come.

Chapter 2

What Is Illegal Fishing, and How Big Is the Problem?

DEFINING ILLEGAL FISHING

Illegal fishing is one of the most complex environmental crimes. The term is enveloped within IUU[1] or illegal, unreported, and unregulated fishing, with the latter pertaining to a wide variety of prohibited acts that are carried out by large and small fishing vessels alike. These activities are either single violations of applicable international, regional, or national fisheries regulations or any combination thereof. IUU fishing activities can occur in internal and archipelagic waters,[2] territorial seas, and 200-nautical-mile exclusive economic zones (EEZs), as well as in the high seas or areas beyond national jurisdictions (see Figure 2.1).

A vessel's fishing activity is considered *illegal* if it (a) uses prohibited gear or methods; (b) operates in closed areas (such as marine-protected areas) or during closed seasons; (c) operates with a fake license or without a license or vessel

Figure 2.1 Maritime Zones (NOAA)

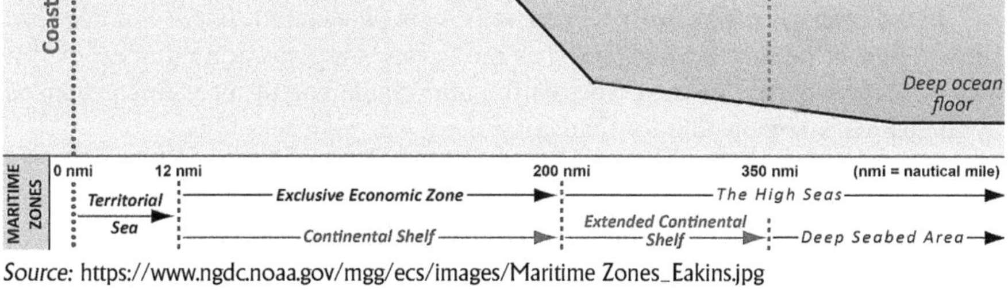

Source: https://www.ngdc.noaa.gov/mgg/ecs/images/Maritime Zones_Eakins.jpg

Box 2.1 *Thunder*: The World's Most Notorious Illegal Fishing Trawler

The 202-foot trawler, *Thunder*, had been banned from fishing in the Antarctic since 2006 because of its numerous violations pertaining to the deployment of illegal nets to fish for Patagonian toothfish, more popularly known as Chilean sea bass. The vessel had allegedly made over $75 million from illegal sales of the fish in the 10 years prior. The trawler, nevertheless, persisted in fishing in this area. The vessel was on Interpol's Purple Notice (similar to a most wanted list) since December 2013. In 2014, the Commission for the Conservation of Antarctic Marine Living Resources, an RFMO that manages the resources in the Antarctic, placed the vessel on its blacklist of fishing vessels. The Sea Shepherd, an environmental group, spotted the trawler poaching in the Antarctic in 2014, and after an epic 110-day, 10,000-nautical-mile ocean chase through the Southern, Indian, and Atlantic Oceans, the trawler's captain sunk the vessel in April 2015 in the waters of São Tomé and Príncipe to, what many suspect, dispose of evidence. The Sea Shepherd rescued the trawler's crew. The captain, Chilean by nationality, was eventually sentenced to three years in prison, and the vessel's crew, which consisted of 30 Indonesian fishers and Spanish officers, was collectively fined $17 million for various violations.[1]

Note

1. Urbina, I. (2015). A renegade trawler, hunted for 10,000 miles by vigilantes. *The New York Times*, July 28.

registration paperwork; (d) operates without a vessel monitoring system; and (e) fishes beyond allocated catch limits (i.e., over quota) or takes juvenile or prohibited fish species.[3] A vessel operates *illegally* in national waters if its activities are either in contravention of the specific country's fisheries laws where it fishes or in violation of laws adopted by the vessel's flag country. For example, in 1981, Indonesia placed a complete ban on the use of trawlers within its EEZ.[4] Therefore, if a trawler, whether foreign or national, fishes in Indonesia's waters, its activity is considered *illegal*. Similarly, an activity can be considered *illegal* in waters managed by regional fisheries management organizations (RFMOs) if it contravenes regulations set forth by these organizations. For example, the European Union (EU) and the International Commission for the Conservation of Atlantic Tunas (an RFMO) banned the use of drift nets in the Mediterranean Sea in 2002 and 2003, respectively, therefore making its use *illegal*[5] in the sea area.

In the high seas, a vessel's fishing activity is considered *illegal* if it does not show a flag or other markings, such as the vessel's registration number, or if it acts in contravention of the conservation and management measures proposed by the RFMOs responsible for managing the given high seas area.

A vessel engages in *unreported* fishing if it fails to report or underreports the catches to relevant authorities (e.g., the country to which it is flagged or where the vessel fishes) or takes prohibited, protected, unauthorized, or endangered species. In the high seas, a vessel's catches are considered *unreported* if it acts in contravention of the management requirements (pertaining to fish and gear) of the RFMO that has regulatory competence in the area where the vessel fishes. For example, fishers are known to engage in the widespread illegal practice of keeping two logbooks: one for the fisheries inspectors, where the catches are underreported, and one "confidential" logbook for the owner, which contains the actual catch.[6] This is considered *unreported* fishing. Unreported fish caught in the high seas that are outside the regulatory reach of RFMOs are not considered *unreported* in the strict sense of the definition of the act.[7]

Lastly, a vessel engages in *unregulated* fishing if it operates in the high seas in contravention of the regulations set forth by relevant RFMO(s).[8] Additionally, an *unregulated* fishing takes place if it is carried out by a vessel in violation of its flag state's international obligations in respect to the high seas or if a vessel catches fish not otherwise managed by the concerned RFMO's conservation rules.[9] *Unregulated* fishing does not apply to the fishing activity carried out inside a country's EEZ, as its legal definition strictly applies to the activities carried out beyond the coastal countries' EEZs. Figure 2.2 is an illustration of these types of IUU fishing.

In addition to these definitions, some RFMOs have specified what illegal fishing means in their areas of competence. For example, the Northeast Atlantic Fisheries Commission, which regulates the fishing activity in the Northeast Atlantic Ocean, considers fishing without a quota or after exhausting quota limits, using prohibited gear or methods, and fishing without a valid authorization

Figure 2.2 Types of Illegal, Unreported, and Unregulated Fishing

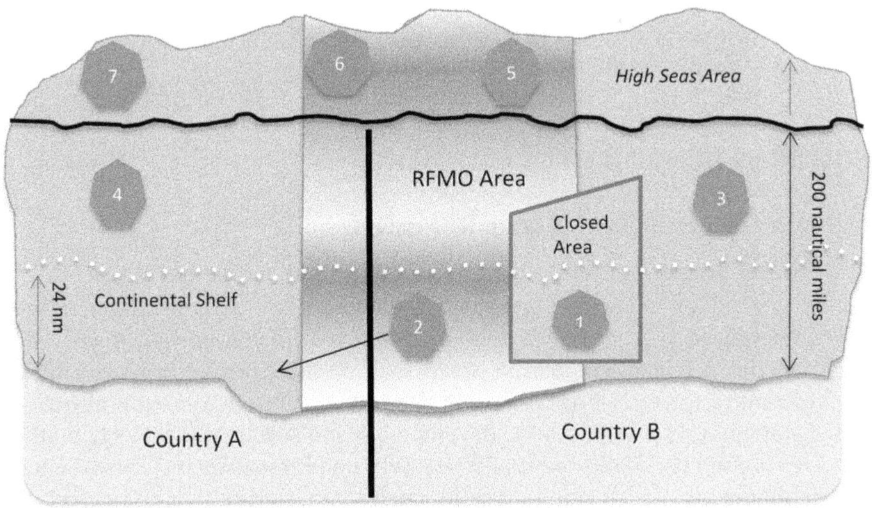

1 = fishing in closed area [illegal]; 2 = 'border hopping' [illegal]; 3 = licensed, but misreporting catches [unreported]; 4 = unlicensed fishing [illegal]; 5 = vessel carrying the flag of a country not party to the RFMO [illegal]; 6 = vessel carrying the flag of the RFMO flag state, but does not comply with the RFMO regulation [illegal]; 7 = unregulated fishing

issued by the flag state member as "serious infringements."[10] The Indian Ocean Tuna Commission, which manages tuna and tuna-like species in the Indian Ocean, lists catching species without a license or vessel registration, violating the conditions of license, fishing without quotas, and fishing with a vessel without a nationality as illegal activities.[11]

As can be seen from the definition of IUU fishing, the act is far more complex than one may expect. IUU fishing does not apply to one single act, and the actual determination of the illegality of the act depends on not only the *modus operandi* (e.g., gear or vessel used) but also the location where the activity takes place. In fact, as simple as the *modus operandi* may sound, it is not as straightforward as it may first seem. Let us recall the EU example on banning the use of drift nets in 2002 in the Mediterranean Sea. The EU was actually unable to successfully enforce this law due to the lack of the definition of what exactly a "drift net" meant. This allowed the French and Italian governments to exploit this major loophole and continue fishing in the area using drift nets. Only in 2007 did the European Commission's Council Regulation No. 809/2007 provide an actual legal definition of what a drift net was, aiming to put an end to this illegal practice in the area. But some suggest that this only put an end to the existing legal loophole, and not the practice per se.[12]

Moreover, one activity may be illegal under some circumstances and not illegal under others, such as in the example of using trawlers in Indonesian waters. Trawlers are not otherwise banned. This is exacerbated by the fact that for some activities, such as those in the high seas or pertaining to some fish stocks, there are no clear guidelines as to what may or may not constitute a prohibited act. This complexity not only makes the enforcement of IUU fishing regulations a colossally difficult task but it also creates opportunities for motivated offenders to readily exploit existing loopholes.

HOW BIG IS THE PROBLEM OF IUU FISHING?

The Calculated Economic Losses to IUU Fishing

IUU fishing occurs in almost all fisheries around the world and is believed to account for a significant proportion of global catches.[13] IUU fishing activities are not confined to any specific type or category of fisheries—such activities are prevalent in inland and marine, small-scale and large-scale fisheries alike.[14] Because more than 2.5 billion people depend on fish for food and nutrition, IUU fishing not only impedes the development and sustainable management of fisheries around the world but it also severely undermines the efforts to reduce global hunger and malnutrition. In some important fisheries, IUU fishing is rampant, striking a significant blow to legal fishing efforts. For example, in 2001, an aerial surveillance conducted over Guinea's EEZ found that, of the 2,313 vessels fishing in the area, about 60 percent were engaged in illegal fishing.[15]

Box 2.2 Seafood Markets

The United States, Japan, and the EU, major global seafood markets, account for about 63 percent of global seafood imports, with the EU importing about 38 percent, the United States about 14 percent, and Japan about 11 percent of the global total.[1]

Recent research has estimated that roughly 30 percent or $2.1 billion worth of wild-caught seafood entering the United States is illegal and unreported,[2] and approximately 50 percent or $1.6 billion worth of seafood sold in the European Union is sourced illegally.[3] This type of behavior directly harms legitimate fishers who sell to the same markets. These fishers also suffer economically due to higher operating costs associated with licensing fees and other compliance measures pertaining to vessel monitoring, catch documentation, or labor laws that need to be observed onboard these vessels.[4] Some studies have estimated that if illegal fishing for tuna is eliminated, legal fishers can make a net income of $18 million over five years.[5]

Notes

1. FAO (Food and Agriculture Organization of the United Nations). (2014). *FAO Yearbook. Fisheries and Aquaculture Statistics*. United Nations. Rome, Italy.
2. Pramod, G., Nakamura, K., Pitcher, T. J. & Delagran, L. (2014). Estimates of illegal and unreported fish in seafood imports to the USA. *Marine Policy, 48*: 102–113.
3. Liddick, D. R. (2011). *Crimes against Nature: Illegal Industries and the Global Environment.* Santa Barbara, CA: Praeger.
4. Ibid.
5. Bonini, S., Saran, N. & Stein, L. (2011). *Design for Sustainable Fisheries—Modeling Fishery Economics*. https://www.mckinsey.com/~/media/mckinsey/industries/social%20sector/our%20insights/saving%20our%20resources%20a%20design%20for%20sustainable%20fisheries/design-for-sustainable-fisheries-modeling-fishery-economics.ashx.

The global fisheries lose an estimated 11–26 million tonnes of fish to IUU fishing every year, making the annual economic loss about $10–23.5 billion.[16] In 2008, a Pew Charitable Trusts study estimated that our seas lose about 108,000 pounds of wild-caught fish *every minute* to IUU fishing.[17] Coastal countries suffer directly from these losses, as only about 10 percent of IUU fishing occurs in the high seas, while the remaining illegal fishing activity occurs in the waters of these countries.[18] Most importantly, developing coastal countries carry the majority of these losses, as an estimated $2–15 billion worth or about 65 percent of their fish resources are removed illegally.[19]

The Geographic Scope of IUU Fishing

Many fisheries scientists have attempted to quantify the geographic scope of IUU fishing around the world. A study examined the illegal fishing activities of vessels that were apprehended between 1980 and 2003.[20] It found that most

Box 2.3 Marine-Protected Areas

Illegal fishing often occurs in marine-protected areas recognized as World Heritage Sites. One such example is the Galápagos National Park—a marine sanctuary that is home to many vulnerable and threatened marine species, such as whales, dolphins, marine iguanas, and penguins. This marine sanctuary also has a great abundance of sharks, thus making it one of the most vulnerable marine-protected areas in the world for illegal fishing of sharks (for their fins) and other fish species, such as tunas.

illegal fishing happened in the Asia-Pacific region, especially in Southeast Asia, and the North and East Pacific Oceans. In 2006, another team of researchers found similar patterns, noting that illegal fishing was especially problematic in China, Indonesia, the Philippines, the Russian Federation, Thailand, and Vietnam.[21]

One study,[22] broadly cited by the scientific community, conducted a global estimate of IUU fishing within the fishing areas designated by the UN Food and Agriculture Organization (FAO). Figure 2.3 uses the data from this study to map these areas and the estimated percentage of IUU fishing that occurred within

Figure 2.3 Percentage of IUU Fishing in FAO Fishing Areas

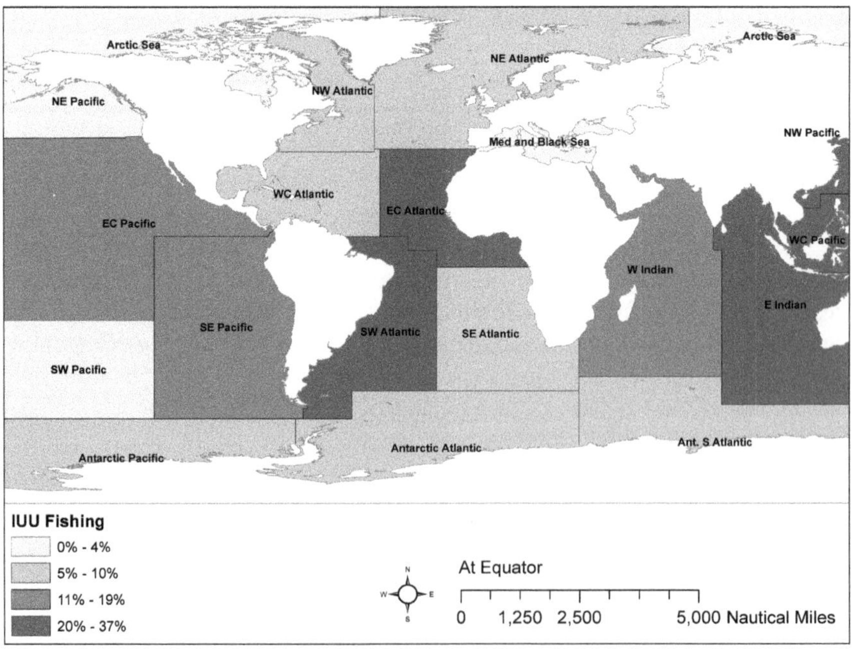

these areas between 2000 and 2003. IUU fishing activities in these areas ranged from as low as three percent (in Northeast Pacific) to as high as 37 percent (in East Central Atlantic). Other vulnerable global fishing areas included West Central and Northwest Pacific, Eastern Indian, and Southwest Atlantic Oceans, where IUU fishing accounted for between 32 percent and 34 percent of all fishing activities in these areas.

Several RFMOs have provided species-level estimates when examining the extent of IUU fishing activities occurring within their management areas. The Commission for the Conservation for Antarctic Marine Living Resources, for example, estimated that in 2003–2004, about 17 percent of the total Patagonian toothfish (*Dissostichus eleginoides*) caught in its management area was illegal, leading to cumulative financial losses of approximately $518 million in 2004.[23] The Northeast Atlantic Fisheries Commission estimated that up to 27 percent of redfish landed in 2002 from its management area was caught by IUU fishing vessels[24]; and the Commission for the Conservation of the Southern Bluefin Tuna estimated that 40 percent of the Patagonian toothfish and 30 percent of the southern bluefin tuna were caught illegally.[25]

Which Species Are at Risk?

Scientists have used different methods to determine which fish species are at a significantly higher risk for IUU fishing. One such attempt was a study conducted by the World Wildlife Fund in 2015.[26] The study evaluated a total of 567 global commercial fish stocks by using different scoring techniques that incorporated both species-specific information, such as value or gear used to catch them, and information about the ocean area where the species are caught. Overall, the study found that 485 (or 85 percent) of the stocks assessed were at a significant risk for IUU fishing, with 304 stocks (or 54 percent) falling within the highest risk category. A great majority of these highest risk species stocks were already overexploited and primarily included high-value and commercially viable stocks[27] (see Table 2.1).

Among the riskiest areas where these species were caught were the Western Central Pacific and the Southeast Pacific regions (refer to Figure 2.3). The researchers found that 90 percent of the commercial species (44 out of 49 species in the Western Central Pacific and 18 out of 20 species in the Southeast Pacific) caught from these two regions were at a high risk for IUU fishing. The same species were assessed to be at a "high" state of exploitation. Table 2.1 lists these high-risk species and the main countries exploiting these species within these two regions.

In the remaining fishing regions, between 66 percent and 88 percent of the assessed commercially harvested species were at a high risk for IUU fishing. These regions also had the same high-risk species, such as banana prawn, Indian mackerel, and largehead hairtail, among others.

Table 2.1 Highest Risk Species Found within Two High-Risk Areas for IUU Fishing

Common Name	Scientific Name	Major Fishing Country
Western Central Pacific		
Bali sardinella	*Sardinella lemuru*	Indonesia
Banana prawn	*Penaeus merguiensis*	Thailand, Australia, Papua New Guinea, Indonesia
Bigeye scad	*Selar crumenophthalmus*	The Philippines, Indonesia, Thailand, Malaysia
Giant tiger prawn	*Penaeus monodon*	The Philippines, Indonesia, Thailand, Australia
Goldstripe sardinella	*Sardinella gibbosa*	Indonesia
Indian mackerel	*Rastrelliger kanagurta*	The Philippines, Indonesia, Fiji
Indian scad	*Decapterus russelli*	Thailand, Malaysia
Largehead hairtail	*Trichiurus lepturus*	Thailand, Singapore, Malaysia
Short mackerel	*Rastrelliger brachysoma*	The Philippines, Indonesia
Yellow croaker	*Larimichthys polyactis*	Taiwan, South Korea, China
Southeast Pacific		
Chub mackerel	*Scomber japonicus*	Ecuador, Peru, Chile
Jumbo flying squid	*Dosidicus gigas*	Japan, China, Chile
Pacific thread herring	*Opisthonema libertate*	Ecuador
Patagonian grenadier	*Macruronus magellanicus*	Chile
Patagonian toothfish	*Dissostichus eleginoides*	Chile
South Pacific hake	*Merluccius gayi*	Peru, Chile
Southern hake	*Merluccius australis*	Chile

Note: The table only lists species that were classified to be in "high" state of exploitation for which specific Latin names were listed. Some species were listed as a group, and the study did not provide any further breakdown at the species level. These are excluded from the list provided in the table.

Two other studies identified specific fish[28] and crustacean[29] species that were at the highest risk for IUU fishing internationally by taking a different scoring approach. These studies used information from available literature and global sustainable English-language seafood guides to learn about the illegally exploited species. These researchers then used a scoring system to assess the risk of these commercially viable species found in these guides and discussed in the literature. Their findings revealed 58 fish and 10 crustacean species that were at a significant risk for IUU fishing. Table 2.2 lists the top 10 fish and all crustacean species identified by these studies.

Table 2.2 Top 10 Illegally Caught Fish and Crustacean Species

Fish		Crustaceans	
Common Name	**Scientific Name**	**Common Name**	**Scientific Name**
Orange roughy	*Hoplostethus atlanticus*	Blue crab	*Callinectes sapidus*
Atlantic bluefin tuna	*Thunnus thynnus*	Deep-sea red crab	*Chaceon quinquedens*
Swordfish	*Xiphias gladius*	Gazami crab	*Portunus trituberculatus*
Atlantic cod	*Gadus morhua*	Golden king crab	*Lithodes aequispinus*
Albacore tuna	*Thunnus alalunga*	Red king crab	*Paralithodes camtschaticus*
Yellowfin tuna	*Thunnus albacares*	Caribbean spiny lobster	*Panulirus argus*
Bigeye tuna	*Thunnus obesus*	Cape rock lobster	*Jasus lalandii*
American plaice	*Hippoglossoides platessoides*	Norway lobster	*Nephrops norvegicus*
Atlantic halibut	*Hippoglossus hippoglossus*	American lobster	*Homarus americanus*
Antarctic toothfish	*Dissostichus mawsoni*	Green spiny lobster	*Panulirus gracilis*

Other species, such as sharks, sturgeons, and totoaba, are among the illegally caught species whose populations have also seen a dramatic drop in only a few decades. One attribute that these species share in common is the demand for their parts rather than their meat. Their parts are more prized than their meat: fins for sharks, raw for sturgeons, and swim bladder for the totoaba.

Box 2.4 The Money Maw, the Aquatic Cocaine, the Totoaba

The totoaba (*Totoaba macdonaldi*) is endemic to the (northern and central) Gulf of California (also known as the Sea of Cortez)—the sea that separates the Baja California Peninsula from the Mexican mainland. This fishery has been the most important fishery in the Gulf of California since the early twentieth century. Adult totoaba are targeted for their swim bladder, which is a highly prized delicacy in China. In China, the swim bladder is consumed as "therapeutic food"[1] and allegedly has a variety of medicinal benefits, including treatments for infertility, blood circulation, skin diseases, and arthritis. It is also believed to be effective in healing lung and kidney diseases. Male fish maw fetches higher prices than female ones, and the prices for a single bladder range from \$8,000 to \$20,000. This price depends on the size, dryness, color, thickness, and oldness of the maw, with the thickest, largest, and oldest being the most prized.[2]

Totoaba swim bladder, commonly referred to as "money maw" or "cash bladder," have also been used as "speculative investment vehicles,"[3] as individuals allegedly buy the bladder when prices drop in hopes to make profit when there is a significant lack of supply of the swim bladder due to the inexorable decline of the fish populations driven by the demand in China.[4] Mexican organized criminal groups are believed to be involved

Figure 2.4 Totoaba Range Map

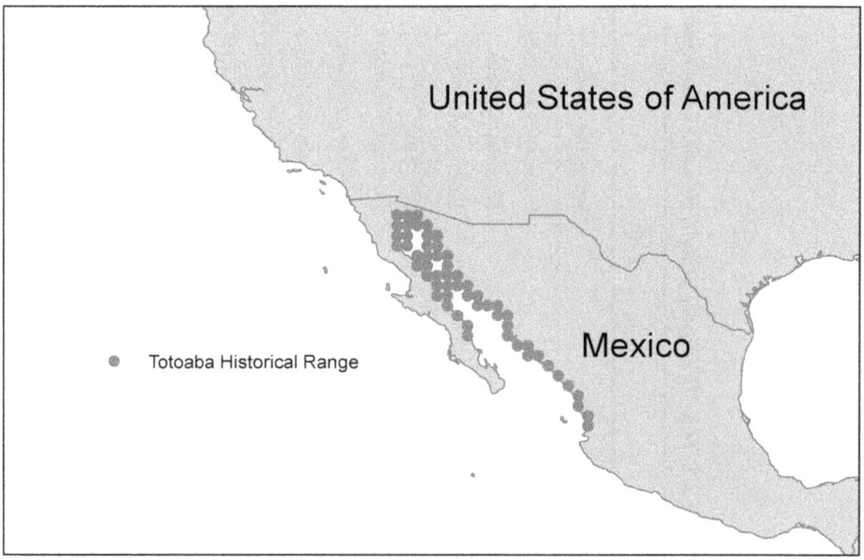

in the international illegal trade of the totoaba, where they make as much money, pound for pound, from this "aquatic cocaine" as they do from cocaine.[5]

The totoaba swim bladder has been exported to China and the Chinese communities in California since the early 1940s. Its reported peak catch was 2,251 tonnes in 1942. However, the unsustainable harvesting of the species led the Mexican government to ban the fishing of the species in the region in 1975, at which time the reported catches of the fish had dropped by about 95 percent.[6] The species is now listed as "critically endangered" by the International Union for Conservation of Nature, and the international trade of the species was banned in 1977. Nevertheless, the illegal fishing of the species persists to this day, threatening its survival, and that of the vaquita dolphin, caught as bycatch at this fishery.

Notes

1. Yagoub, M. (2016). *"Narco Fish" Trafficking Easy Money for Mexico Criminals.* Accessed November 2016, https://www.insightcrime.org/news/brief/narco-fish-trafficking-easy-money-for-mexico-criminals/.

2. Greenpeace. (2015). *How the Illegal Totoaba Trade in Hong Kong Is Pushing Mexico's Vaquita to the Brink of Extinction.* https://www.greenpeace.org/eastasia/Global/eastasia/publications/campaigns/Oceans/HK%20Totoaba%20Trade_Greenpeace%20Media%20Briefing.pdf.

3. Hofford, A. (2010). *Yet More Endangered Species on Sale in Hong Kong, This Time the "Larke Croaker" Fish.* Accessed December 2016, http://www.alexhoffordphotography.com/node/2303.

4. Greenpeace. (2015). *How the Illegal Totoaba Trade in Hong Kong Is Pushing Mexico's Vaquita to the Brink of Extinction.* https://www.greenpeace.org/eastasia/Global/eastasia/publications/campaigns/Oceans/HK%20Totoaba%20Trade_Greenpeace%20Media%20Briefing.pdf.

5. Lah, K. & Moya, A. (2016). "Aquatic cocaine": Fish bladders are latest Mexican smuggling commodity. *CNN Health.* http://www.cnn.com/2016/05/23/health/aquatic-cocaine-totoaba-bass-smuggling/index.html.

6. United Nations Fisheries and Aquaculture Department (2015). *FAO Fishery Statistics, FAO FishStat Database.* http://www.fao.org/fishery/statistics/en.

Who Are the Offenders?

Is illegal fishing an opportunistic undertaking, or is it an organized criminal activity? This question was raised for the first time in 2008 during the ninth meeting of the United Nations Open-ended Informal Consultative Process on Oceans and the Law of the Sea (UNICPLOS) and at the meeting of the Conference of the Parties to the UN Convention against Transnational Organized Crime.[30] As it turned out, this was not an easy question to answer, as even the international experts present at these meetings had divergent opinions about whether there is enough "organized" activity in the stricter sense of the definition of organized crime that can be used to make this potential link. Nevertheless, the discussions at these meetings created a forum for a larger international inquiry and academic assessment of the potential relationship between organized crime and IUU fishing.

For some important fisheries, such as the abalone fishery in South Africa, finding a genuine link between organized crime and IUU fishing is not difficult. The illegal harvesting of abalone in the country increased in the mid-1990s, which coincided with China's economic growth and subsequent increasing demand for abalone in Hong Kong and Taiwan.[31] Asian criminal enterprises expanded in South African abalone fisheries in part due to China's economic development and were driven by the high profits that could be made of the species in Asian markets. Abalone was worth up to $4,000 per kilogram in some prime Asian markets. Known Chinese organized criminal groups, such as the Triads, had already been present in South Africa since the early 1920s.[32] Not surprisingly, the Triads saw abalone as yet another resource to be exploited when its demand skyrocketed in the 1990s. The poaching of abalone in South Africa is quite often associated with violent encounters and activities of sophisticated criminal enterprises[33] that also include such local gangs as South Africa's Cape Flats.[34]

Organized criminal groups also target important fisheries in Australia, including the abalone, rock lobster, and shark fisheries. A national survey of fisheries officers and stakeholders in the country found that abalone was the most vulnerable fishery to fall prey to local outlaw motorcycle gangs and Chinese organized crime groups. These gangs have used various strategies to avoid detection, which included modifying transport vehicles and fishing boats to create concealed spaces for laundering the species, countering surveillance by using lookouts or guards, working in crews to facilitate distribution, or engaging in documentation fraud.[35]

The Russian mafia is involved in the harvesting and sale of beluga sturgeon in the Caspian Sea.[36] In the 1990s, these criminal syndicates were estimated to earn an annual income of $4 billion through the illegal export of the Caspian Sea sturgeon into the United States, the EU, and Japan.[37] Considering that about 90 percent of the global trade in caviar comes from the Caspian Sea,[38] it is not surprising to see the Russian organized criminal groups' involvement in this lucrative trade. These criminal groups recruit local fishers to catch sturgeon. These fishers, who work in groups of three to four people, are guaranteed their protection or *krysha* (literally "roof"). The Russian mafia provides these fishers with either *kriminal'naya krysha* (protection by criminal organization) or *militseiskaya krysha* (protection by the police).[39] The first type of protection may come in the form of provision of weaponry, such as hand grenades and automatic firearms, to be used for self-protection in case of confrontation with the police.[40] The second type of protection is guaranteed if they have initially bribed the police to not inspect their marked vessels.[41]

Whatever the degree of involvement of organized criminal groups is, scholars have argued that it is best to treat illegal fishing as a "serious crime" and organized criminal activity rather than a fisheries management problem.[42] IUU fishing is a well-organized activity that escalates into a foreign threat by breaching countries' sovereignty and involving a systematic theft of their resources.[43] Such activities are often carried out in complete disregard for regulations, involve a complex network of operators, and have long-term economic and social consequences. Treating illegal fishing as an organized criminal activity would mean that the international legislative instruments available for dealing with other transnational organized crimes can be used to deal with this crime more effectively. In fact, it is not difficult to imagine the attractiveness of IUU fishing to organized criminal groups. First, the chances of detecting such activity in the vastness of the oceans are low. Second, illegally caught fish can easily be transformed into another commodity or concealed with legally caught fish. Third, illegally caught fish can virtually be untraceable. Last, fishing is an extremely high-profit, low-cost activity, and is, therefore, an ideal capital investment.[44]

What Are the Economic and Social Impacts of IUU Fishing on Developing Countries?

> In Africa's coastal waters IUU fishing has reached epidemic proportions. This plunder destroys entire coastal communities when they lose the opportunities to catch, process and trade. Coastal trawlers . . . are engaging in organized theft disguised as commerce.[45]
>
> —Kofi Annan
> Former UN Secretary General and Chair of the Africa Progress Panel

Economic Impacts

Due to depleting resources in their EEZs, rich fishing countries are turning to developing states for their marine resources. Wealthier nations, driven by market pressures and taking advantage of the inequalities between them and developing nations, have found ways to exploit these developing countries' marine resources on which these people depend for food and livelihood.[46] An example of this is the EU's cooperation agreements with West African coastal countries. The high demand in the world's largest market for seafood products,[47] the EU, far outstrips the reproductive capacity of the fisheries within the EU waters. Thus, the EU is actively negotiating trade agreements with West African countries to access their resources. These access agreements, while may be seen as ways to reduce poverty and promote sustainable development in these poor countries, are, in fact, purely commercial arrangements designed to benefit primarily the EU countries.[48] These cooperation arrangements affect the development of sustainable fisheries in these West African countries, all the while providing employment opportunities for fishing fleets and related sectors of the EU member states, primarily benefiting France, Spain, Italy, and Portugal. Meanwhile, Guinea-Bissau, a country that has multiple fishing agreements with the EU, despite its wealth of fish stocks,[49] is among the poorest coastal countries in Africa.

This demand has brought increasing external pressures on the developing countries, which, until recently, due to the paucity of expertise, lack of technology, and inadequacy of enforcement in their waters, have not been able to sufficiently manage their marine resources.[50] In some of these countries, the situation is more difficult to handle due to the corruption of fisheries officials.[51] This inability to monitor and control the access of foreign fishing fleets into their EEZs has not only led to depleting their fish stocks but also significantly affected these poor countries' artisanal fisheries. For example, in Cameroon, artisanal fishers have reported purposeful and malicious destruction of their gear by illegal trawlers from China.[52] Semi-artisanal shrimp fishers in Guinea, Sierra Leone, and Liberia have suffered from direct (e.g., industrial IUU vessels running them down) and indirect (e.g., same IUU vessels removing all available fish or shrimp) impacts.[53] Illegal fishing vessels in West Africa also removed about 35 percent of the income that artisanal fishers could have otherwise earned.

Some equate this type of opportunistic behavior to "stealing food from the poorest of the world"[54]—a theft that reaches $9 billion a year in direct losses suffered by developing countries, of which $1 billion is from Africa alone.[55] Such behavior not only takes away the food supply of poor fishing communities along the coasts of developing countries but also often results in their displacement and substantial reduction in income, exacerbating their already destitute living conditions. Even more troubling is the fact that this "theft" of resources can be carried out via private or direct agreements, which are signed between individual vessel owners and the governments of the coastal countries, where the foreign fishing vessel carries no accountability for reporting to its flag state. For example, a recent study conducted by Oceana, an environmental nongovernmental organization (NGO), discovered that between 2012 and 2016, a total of 19 EU vessels flagged to Greece, Italy, Portugal, and Spain spent 31,000 hours fishing in the waters of Gambia and Equatorial Guinea under private agreements that were in direct contravention of EU laws.[56]

Coastal countries in the Asia-Pacific region experience similar problems. These countries lose about $5.8 billion a year to IUU fishing, accounting for about 8–16 percent of the total annual reported catch.[57] For example, Indonesia, which is one of the most important fishing countries in the world, loses about $4 billion a year in profit due to illegal fishing.[58]

A recent study of the world's three major fish markets—namely, the EU, the United States, and Japan, revealed that in recent years, the EU has increased its bilateral fishing deals with West Africa, East Africa, and a few countries in the South Pacific (Kiribati, Solomon Islands, and Micronesia). Over 60 percent of Japan's marine catches are from its neighboring EEZs. Lastly, U.S. fish consumption has reached most of the world's fisheries, particularly those along the coastlines of South America and Southeast Asian.[59] Not surprisingly, the EU and the United States are estimated to consume about $1.6 billion[60] and upward of $2.1 billion,[61] respectively, in illegal fish products every year.

Social Impacts

The social impacts suffered by these developing countries are unquantifiable. To increase revenue by avoiding running costs, illegal fishing vessel operators recruit crews from developing countries, as this allows them to take advantage of largely unregulated labor markets and virtually no controls on working conditions.[62] These workers are subjected to various human rights abuses aboard the vessels, including working in dangerous conditions and imprisonment without food and water, as well as physical and verbal abuses.[63] They are "*de facto* prisoners of the sea," who are forced to work more than 18 hours a day and are regularly subjected to severe physical and sexual abuses.[64] One such account involves a Cambodian fisher who was trafficked into a Thai-flagged trawler and was forcefully kept onboard the vessel for over two years shackled with a rusty metal collar around his neck, only to catch pet food and livestock feed.[65] A study conducted by the UN Office on Drugs and Crime found that about 60 percent of Cambodian fishers trafficked to work in the Thai fishing industry reported that they had witnessed a ship's captain murdering a coworker.[66]

Children are also recruited or trafficked in the fishing industry. Many young men working onboard fishing vessels have, in fact, started their career as child laborers. A fact-finding mission to Gwadar district, Pakistan, revealed the following disturbing facts about child exploitation onboard these vessels:

> Owners hire children for Rs. 1,500–1,800 [$17–21] per month and take them into the deep sea for two to three weeks. Once at sea, these children without any parental supervision and protection are at risk of being subjected to physical, emotional and sexual abuse. Interviews with some former Walaed revealed that in the camouflage of helping the cook, the children (Walaed) are sexually exploited by the adults on board. On some occasions, 5–8 adults may have sex with a single child during the fishing trips, exposing them to the risks of genital trauma and sexually transmitted infections, including HIV/AIDS.[67]

Most frequently, these trafficked children, estimated to be in thousands, are recruited from local poverty-stricken communities where there are fewer, if any, job prospects. These children are then subjected to not only brutal beatings and abuse onboard these vessels but they are also exposed to life-threatening experiences. Children recruited in Senegalese purse seine fisheries, for example, work as divers to guide fish into the nets, exposing them to the danger of drowning.[68]

What Are the Environmental Impacts of IUU Fishing?

Destructive Fishing Methods

Illegal fishing vessels often use destructive fishing practices, such as bottom trawling, blast fishing, poison fishing, cyanide fishing, drift nets, and muro-ami nets. Using these methods obliterates the key components of the marine ecosystem, often leaving a devastating and permanent damage on the oceans.[69]

Bottom trawls, which are among the most destructive gears in use,[70] are heavy and wide nets with weighted metal rollers that are dragged across the seafloor. These trawls can destroy large areas of seafloor habitats because these nets crush everything in their path, such as coral formations, and kill tonnes of fish and other animals inhabiting the bottom of the ocean. Losing corals is detrimental for the oceans, as they are considered fundamental building blocks of the marine ecosystem and play a crucial role in it. These reefs are also among the most endangered habitats on the earth. Coral reefs are home to over one million marine species that use these reefs as nursery, refuge, or feeding areas. Coral reefs also play an important role as natural barriers that protect the coastal areas by minimizing the waves and cyclones on these areas, as well as by protecting these areas from storm surges.[71] These reefs already suffer from a multitude of other threats, such as bleaching due to elevated sea surface temperatures, and other local-scale human disturbances, such as coral mining, agricultural and urban runoffs, and introduction of invasive species.[72] Therefore, bottom trawling, and, indeed, other destructive fishing methods, adds more pressures to these already distressed reefs.

Blast fishing involves the use of a bomb or dynamite set to explode under water. These explosions send shock waves into the water, rupturing the swim bladder of the fish (the organ that controls their buoyancy). As a result, fish float to the surface of the water to be collected with seine nets, while other damaged

Box 2.5 Telephoning Fish?

Many illegal fishers have been using innovative techniques to capture fish from the wild. One such example is the method known as "telephoning fish," which is actually a method of electrocuting fish to stun them to the surface of the water. Illegal fishers in the United States "telephone fish" by using rotary telephones that require cranking a handle. They then place the charged wires that transmit the charge into small bodies of water and turn the crank to generate a strong enough electrical current to stun the nearby fish. The electrocuted fish that are temporarily paralyzed float to the surface of the water to be retrieved.[1] A similar technique, referred to as "pulse fishing," is used in the Netherlands by trawlers. These vessels use nets that generate an electric current to stun fish, especially sole, which, when electrocuted, jerk upward, making it easier for trawlers to catch them.[2]

Notes

1. Green, E. K. (2011). Telephoning fish: An examination of the creative deviance used by wildlife violators in the United States. *International Journal of Rural Criminology, 1*(1): 23–39.
2. Fleming, A. (2018). Pulse fishing: Is it OK to "electrocute" fish? *BBC News*, January 16. http://www.bbc.com/news/world-europe-42692924.

fauna (such as juvenile fish, fish larvae and eggs, and corals) sink to the bottom of the ocean.[73] In some parts of the world, between 20 and 50 blasts are heard every day.[74] This practice is used in over 30 countries. In Southeast Asia, it is most frequently used in Malaysia, the Philippines, and Indonesia, while in Africa, its use is rampant in Tanzania.[75] This practice has led to the loss of over 50 percent of the coral reef system in Southeast Asia[76]—an impact that some scientists have estimated would require about 100 years to recover from.[77] Additionally, blast fishing has reduced the productive capacity of coral reefs to one-fifth of their original capacity.[78] Blast fishing also significantly reduces demersal plankton that many reef fish feed on, thus affecting the marine food web.

Poison fishing is fishing with the aid of poisonous plants or substances. In Africa, for example, the fishers sew the plants into the water, and within a period of time, which varies according to conditions, stunned fish rise to the surface of the water and can be taken by hand. There are about 325 such plants in Africa.[79] In China and Taiwan, fishers derive rotenone—a traditional fish poison—from derris root. This plant-based poison is used to impede the ability of fish to take up oxygen, as a result of which the fish become narcotized and easy to collect.[80] *Cyanide* is among the most widely used toxic substances used for fishing. This poison kills coral polyps, and this damage leads to the discoloration of coral colonies.[81] The method was first introduced in the Philippines in the 1960s and was used to primarily capture reef fish and aquarium species. The poison not only stuns the target species, which then float to the surface of the water, but it also kills the corals and more vulnerable small fish and invertebrates in that area.[82]

Drift nets, also nicknamed "walls of death,"[83] are synthetic gillnets that extend about 15–35 miles and float in surface or mid-waters to catch salmon, tuna, billfish, swordfish, and other species. These are vertical sheets of nets that

---

Box 2.6 Buckeye Tree Leaves and Walnut Hulls?

North American buckeye tree leaves and walnut hulls have been used by illegal fishers in the United States to stun the fish to the surface of the water. As one poacher put it, "You know walnuts have all that dark stuff. If you put a bunch of walnuts in a sack after you smash them up and then put it in a stream, it will take the oxygen out of the water. By taking out the oxygen, it makes the fish immediately downstream float to the top. If there was enough, it could kill them, but usually it just knocks them out." These methods were known to and passed down from Native Americans who used the ground buckeye leaves as a powder for the exact same purpose.[1]

Note

1. Green, E. K. (2011). Telephoning fish: An examination of the creative deviance used by wildlife violators in the United States. *International Journal of Rural Criminology, 1*(1): 23–39.

---

are held in place by attaching weight-lines to the bottom of the net. These free-swinging meshes entangle or gill the fish that swim into it. Importantly, these nets entangle a significant number of nontarget species that are later discarded, such as sea turtles, seals, dolphins, porpoises, whales, sharks, and albatrosses,[84] thus impacting the populations of these species.[85]

Other illegal fishing methods, widely used in Southeast Asia, are *muro-ami nets*. These are nets with weighted bags that contain either heavy stones or large cement blocks that are pounded to startle fish out of crevices and corals into the net.[86] The damage on the coral reefs caused by this method is often irreparable. These nets also affect the nontarget species.

The Problem of Bycatch

IUU fishing already leaves a significant ecological impact on the species that are targeted for commercial purposes. However, these species are not the only ones being affected by it. Many more species that suffer as a result of IUU fishing are the incidentally caught nontarget species that are thrown back into the oceans. These are collectively called "bycatch" and include a variety of fish, mammal, reptile, and bird species. Bycatch is a significant problem for several reasons. First, the removal of nontarget species disrupts the food supply of larger, predatory species that depend on these species as essential sources of food. Second, the species, when dumped into the oceans dead, putrefy, and this affects the seabeds.

The problem of "bycatch" was first brought to public attention in the late 1960s and early 1970s, following the media outcry about the common, spinner, and striped dolphins that were killed by purse seines set in the Eastern Tropical Pacific tuna fishery.[87] Dolphins were captured in those nets because tuna often swam right beneath schools of dolphin. As tuna was hauled, several hundred thousand dolphins drowned in the process. The public outcry following these revelations over the deaths of dolphins led not only to the mandatory practice of using the "back down" method when hauling tuna but also to the mandatory labeling of tuna as "dolphin-friendly" to attest to its sustainable catch.[88]

As much as 25 percent of all marine species caught in global fisheries are thrown back into the sea because they are not the intended target and are often treated as waste. This annual "waste" quantifies into approximately 300,000 whales, dolphins, and tortoises; 100 million sharks; 480,000 metric tonnes of shrimp[89]; and 100,000 seabirds.[90] Many of these nontarget species are already critically endangered. Shrimp trawlers account for the highest rate of "bycatch" within the seas,[91] where an estimated 35–85 tonnes of "waste" is discarded for every 100 tonnes of shrimp caught.[92] Long-line fishing vessels that set several thousand baited hooks to catch tuna and Patagonian toothfish have had a dramatic impact on many already critically endangered albatross and other seabird species. These birds strike at these baited hooks, become entangled, are dragged

Box 2.7 Vaquita: The World's Most Critically Endangered Marine Mammal

The vaquita, also called the "panda of the sea," is the world's smallest porpoise and is critically endangered. The species is only found in the Gulf of California, same as the totoaba. The vaquita, with an estimated population of 30 (of which only an estimated 8

Figure 2.5 Two Vaquitas (Paula Olson, NOAA)

Figure 2.6 Vaquita Range Map

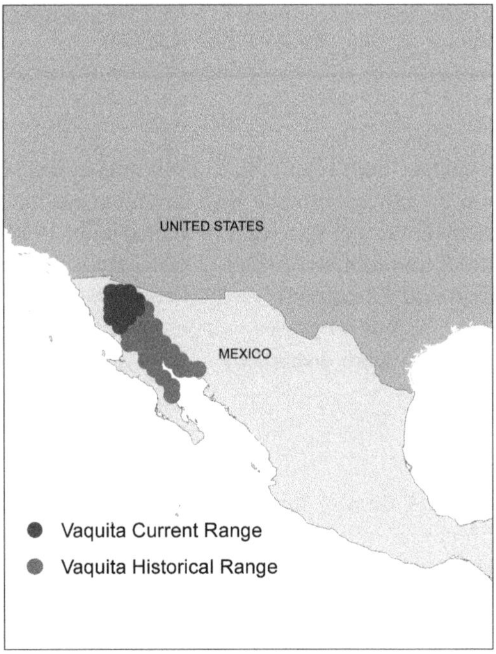

are breeding females[1]) as of 2017, faces imminent extinction, because it is bycatch in the fisheries that use gillnets to catch the totoaba.[2] The vaquita is trapped in the gillnets that remain in the waters for days or weeks at a time to catch the totoaba because these nets are not visible to the porpoise in the water. Without the chance to surface for air, which the vaquita needs in order to survive, the porpoise drowns.

The Mexican government, with help from the United States, is taking some dramatic steps to save the vaquita. The VaquitaCPR marine conservation organization in Mexico is placing autonomous acoustic detectors in water columns to help them detect, rescue, and temporarily relocate the vaquita into sea pens or ocean sanctuaries.[3] The U.S. Navy-trained bottlenose dolphins that are already equipped with the incredible natural sonar abilities have been deployed in 2017 to work as "underwater detectives" and help VaquitaCPR, and their job is to track the remaining vaquitas and lead them to safety.[4] However, these methods may not be working as was hoped as vaquita numbers keep dropping, and the vaquita will probably go extinct.

Notes

1. Mosbergen, D. (2017). U.S. Navy dolphins join the fight to save endangered vaquita porpoises. *Huffington Post*, January 4. https://www.huffingtonpost.com/entry/dolphins-vaquita-porpoise_us_586c9901e4b0d9a5945d2801.
2. TRAFFIC Bulletin. (2014, October 29). *Wildlife Watch: Strengthening Capacities for the Protection of Nature in Mexico*. Accessed November 29, 2016, http://www.traffic.org/home/2014/10/29/wildlife-watch-strengthening-capacities-for-the-protection-o.html.
3. VaquitaCPR. (2017). *How Does the Acoustic Detection of the Vaquita Work?* https://www.vaquitacpr.org/how-does-the-acoustic-detection-of-the-vaquita-work/.
4. Mosbergen, D. (2017). U.S. Navy dolphins join the fight to save endangered vaquita porpoises. *Huffington Post*, January 4. https://www.huffingtonpost.com/entry/dolphins-vaquita-porpoise_us_586c9901e4b0d9a5945d2801.

---

underwater, and drown as these lines sink.[93] Drift nets in the Mediterranean Sea, used in the past to primarily catch swordfish and albacore tuna, have had a dramatic impact on many nontarget species. For example, in 1990–1991, the Italian fleets used these nets to catch swordfish. Their catch, however, consisted of 18 percent swordfish and 82 percent bycatch comprising 85 different species, such as turtles, sharks, harbor porpoises, whales, juvenile bluefin tunas, and dolphins, almost all of which were discarded.[94]

The Problem of Discards

High-value target species are also sometimes discarded. This is primarily due to the quota system imposed on vessels by fisheries management bodies. When a vessel meets the quota for one high-value species, fishers may move on to another high-value fish and discard the one for which they have reached the quota. Similarly, vessels that operate on quota are selective in the specimens they keep, retaining only the largest and discarding the rest.[95] This practice is known as "high grading," and it not only contributes to the depletion of stocks but it also affects stock assessments.[96]

Serial Depletion

Large, slow-growing predatory fish, such as cod, halibut, and grouper, and high-value invertebrates, such as shrimp, lobster, and large shellfish, are particularly affected by illegal fishing.[97] In many regions of the world, illegal fishing has led to "fishing down the food web," a phrase coined by marine biologists to explain this serial depletion of species. The scientific phrase refers to the process where marine biodiversity is significantly reduced when short-lived, fast-growing, and small fish gradually replace large predatory species.[98] This may have far-reaching consequences for the functioning of the marine ecosystem. For example, the decline of great sharks in the Northwestern Atlantic "has triggered a tropic cascade that collapsed a century-old fishery for bay scallops."[99]

In other marine habitats, such as kelp forests, coral reefs, and estuaries, this "fishing down the food web" may lead to other ecological problems. The increased outburst of the previously suppressed smaller species does not necessarily mean humans will have a different pool of marine species to target. It means that "fishing down the food web" will reduce the "number and length of pathways linking food fishes to the primary producers, and hence [leading to] a simplification of the food webs."[100] A diversified food web, where different types of prey are available in case of population fluctuations of other prey, allows for coexistence of the species in the oceans, dependence on each other, and adjustment in case of ecological impacts.[101] A more simplified food web would mean the larger predatory species will have much fewer prey to depend on,[102] thus significantly pressuring their own populations and putting them at a high risk for population fluctuations. Figure 2.7, adopted from a research published in *Science* magazine in 2001,[103] illustrates two simplified top-down food web interactions between species in kelp forests in Alaska and California and Gulf of Maine before and after overfishing.

CONCLUDING REMARKS

Dealing with IUU fishing is a gargantuan task, as, by its very definition, the crime cannot be easily classified. Each component of I-U-U is a distinct type of crime and describes activities that depend on the characteristics of the actors involved (e.g., unlicensed fishers chasing fish that can be legally caught), the gear and the location (e.g., using prohibited drift nets to fish in the Mediterranean Sea), and the species targeted, among others, to be considered illegal.

Figure 2.7 Food Web Interactions in Kelp Forests in Alaska/California and Gulf of Maine before and after Overfishing

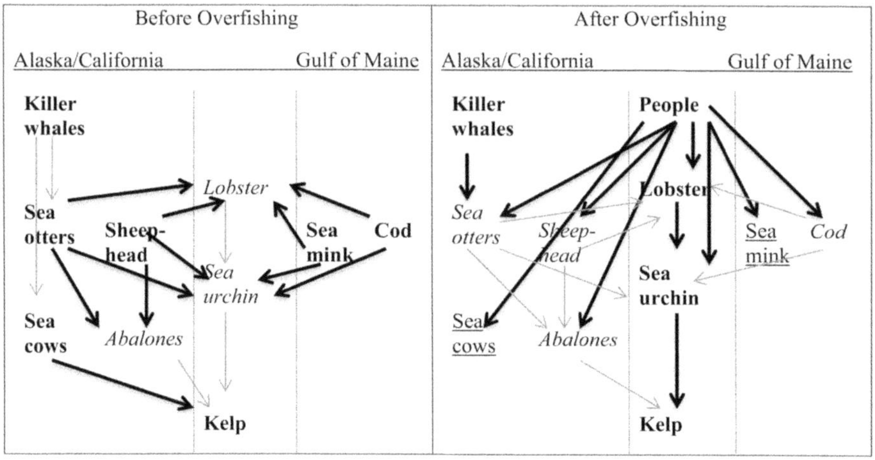

Bold = abundant; Italic = rare; Underlined = extinct. Thick arrows = strong interactions; thin arrows = weak interactions

Nevertheless, IUU fishing remains a complex crime with far-reaching social and environmental consequences. While every coastal country suffers directly from illegal fishing, developing nations are among the ones most affected by it. This is primarily due to the growing global demand for fish and fish products, as well as increasing dependence on these countries to supply the major global markets with fish and fish products. To carry out their illegal activities, fishers engage in various human rights abuses and exploitations of not only adult but also child fishers who are recruited from these developing countries. These fishers, who are often linked to organized criminal groups, do not hesitate to use violent methods to carry out their illegal activities. Neither do these illegal fishers reflect on the environmental consequences of their destructive actions (such as the use of harmful gear and poisonous substances to capture the target species), which not only directly affect the already depleted target species but also deprive vast expanses of the oceans of the vital ecological balance necessary for a healthy ecosystem to thrive.

Chapter 3

Why Do People Fish Illegally?

WHAT DO THE COUNTRIES WHERE ILLEGAL FISHING IS WIDESPREAD SHARE IN COMMON?

Illegal Fishers Use Flags-of-Convenience to Fish in Their Waters

International maritime law requires that every merchant ship be registered in a country, that is, have a flag country. No vessel can leave or arrive into a port without flying a flag. Accordingly, when a vessel is registered to a state, it is subject to that state's regulatory control and, therefore, operates under its laws.[1] Meanwhile, these countries are responsible for maintaining a register to keep track of ships that fly its flag, so that they can exercise jurisdiction over these vessels and their crew should legal issues arise. These countries are also responsible for ensuring the safety of navigation, seaworthiness of the ship, labor conditions, and all other "technical and social matters" related to the vessel.[2] Importantly, these countries have to ensure that the vessel carrying its flag complies with not only domestic but also international maritime laws.

When ships fly a flag in a sovereign state that is different from that of the shipowner, it operates a so-called flag of convenience (FOC). This practice became popular after World War I when nonmaritime countries, such as Panama, Liberia, and Honduras, began registering foreign-owned vessels for economic reasons.[3] However, this practice has historical precedent going back to at least the sixteenth century. In the sixteenth century, due to the West Indies trade restrictions, British vessels mostly used Spanish flags. They also used the French flag on fishing vessels from Newfoundland in the seventeenth century as a means to avoid restrictions imposed by their country.[4] In the nineteenth century, the use of FOCs was common among those who evaded antislavery agreements.[5] American vessel owners, frustrated with increasing regulations and growing labor costs in the United States, started registering their ships in Panama and have been doing so since the 1920s.[6] In the Prohibition era, the American vessel owners used Panamanian flags on their ships to serve liquor on board.[7] The use of FOCs by the fishing industry, however, has proliferated in the past two decades,[8] making the dealing with illegal, unreported, and unregulated (IUU) fishing a challenging task.

Box 3.1 Vanuatu

Vanuatu is a small island country made up of approximately 80 islands that stretch 1,300 kilometers in the South Pacific Ocean. It is also one of the known FOC states that has been registering a significant number of vessels with a history of IUU fishing. The country has continuously withstood the pressure from countries such as Australia and New Zealand and many nongovernmental organizations to tighten its controls and disclose the identities of the owners of these vessels. As of 2016, about 94 percent of the country's fishing vessels were foreign ships mainly from Japan, Poland, Russia, Canada, and Greece,[1] and a great majority of these included tuna freezing and transporting vessels.

Note

1. Vanuatu Maritime Service Limited. (2016). *Vanuatu Maritime Ships—Vanuatu International Ship Registry*. New York, NY. http://www.vanuatumaritimeships.com, as cited in Daniels, A., Gutiérrez, M., Fanjul, G., Guereña, A., Matheson, I. & Watkins, K. (2016). *Western Africa's Missing Fish: The Impacts of Illegal, Unreported and Unregulated Fishing and Under-Reporting Catches by Foreign Fleets*. Overseas Development Institute.

States that operate "open registries" or allow foreign-owned vessels to fly their flag are often referred to as "flags-of-convenience states" (FOCSs). Many of these FOCSs do not have the resources or are unwilling to monitor and control vessels flying their flag. These same states impede efforts made by regional fisheries management organizations (RFMOs) to deal with IUU fishing. Any effort to board and check for violations of fisheries management regulations in the high seas imposed by the RFMO has to be approved by the flag state. Many of these FOCSs, however, prohibit these RFMOs to board, inspect, or penalize their vessels.[9] Importantly, a state can only be bound to legal requirements if it has ratified the pertinent international fisheries instruments, and many of these states operating "open registries" have not done so.[10] Lastly, many of these "open registries," despite their international obligations, do not investigate or take into account whether or not a fishing vessel has had a history of illegal fishing before registering them under their flag.[11] Consequently, illegal fishing vessels are increasingly using FOCs because this allows them to bypass international and national fisheries regulations and controls, as well as to avoid penalties when and if caught.

FOCs are easy and quick to acquire and can be obtained over the Internet for a few hundred dollars. Vessel owners use online sites, such as www.flagsofconvenience.com, to register their ships.[12] In fact, private companies operate some open registries, such as the one above, and these companies are typically based in countries different from the flag state. The companies approach developing countries with proposals to operate an open registry on their behalf and work on a commission basis. They make the registration process fast and easy to further incentivize fishing vessels to register under FOCs.[13]

Box 3.2 Anarchy at Sea

"No one pretends that a ship comes from the home port painted on its stern, or that it has ever been anywhere near. Panama is the largest maritime nation on earth, followed by . . . Liberia, which hardly exists. No coastline is required either. There are ships that hail from La Paz, in landlocked Bolivia. There are ships that hail from the Mongolian desert. . . . "Liberia" is run by a company in Virginia, "Cambodia" by another in South Korea, and . . . "Bahamas" by a group in the City of London. . . . By shopping globally, shipowners found that they could choose the laws that were applied to them rather than haplessly submitting as ordinary citizens must to the arbitrary jurisdictions of their native states."[1]

Note

1. Langewiesche, W. (2003). Anarchy at sea. *The Atlantic Monthly*, September.

Vessels reflag and change names several times in a season, which is a practice known as "flag-hopping," and do so remotely,[14] sometimes by fax while at sea at a moment's notice and at a very little cost.[15] It is also common practice among IUU fishing vessels to leave the port with one flag and change its flag to an FOC when fishing illegally in an area.[16] The name, nationality, and country of residence of the true owners of these vessels are often carefully hidden,[17] which makes it extremely difficult to identify and penalize the illegal vessel owners.

Panama, Belize, Honduras, and St. Vincent and the Grenadines top the list of 10 FOCSs in terms of numbers of large-scale (greater than or equal to 24 meters in length) fishing vessels registered under their flags. The top 10 FOCSs account for 95 percent of all such registered vessels (see Table 3.1). RFMOs consider these countries as "flag states of particular concern"[18] because historically they have shown little interest in dealing with IUU fishing. In fact, vessels carrying their flags can operate with total disregard for any RFMO regulations, because these FOC countries are not RFMO members and, as such, are not bound by their management regulations.[19] Meanwhile, more than 75 percent of large vessels fishing internationally fish under the flags of the top four countries,[20] thus frustrating the efforts to effectively deal with the IUU fishing problem globally. Some estimates suggest that approximately 15 percent of the world's large fishing vessels, including 1,000 vessels in the largest category, use FOCs to carry out their activities.[21] This allows the beneficial owners of the vessels to enjoy ultimate anonymity, and the ability of using FOCs is the exact reason why many of the fishing vessels engage in such practices.[22]

Several researchers have identified the beneficiary countries of these FOCs. The past and present studies suggest that the EU countries, especially Spain, and countries in East Asia dominate the ownership of FOC vessels. The most recent study conducted in 2014 analyzed fishing vessels registered in

Table 3.1 Percentage of FOC Vessels Registered to Top FOC States in 2005

Flag State	Percentage of All FOC Vessels	Cumulative Percentage
Honduras	32.8	32.8
Belize	19.0	51.8
Panama	17.5	69.3
St. Vincent and the Grenadines	5.8	75.1
Georgia	4.7	79.8
Vanuatu	3.7	83.5
Cambodia	3.7	87.2
Equatorial Guinea	3.1	90.3
Cyprus	2.1	92.4
Sierra Leone	2.1	94.5
All others	5.5	100.0

Note: Data for this table come from Gianni & Simpson (2005).

www.seaweb.org and discovered that, for example, Spain accounted for 21 percent of the large-scale vessels registered under an FOC. Table 3.2 summarizes the findings derived from this study.[23]

Table 3.2 List of Top 10 Fishing Countries and Their Large-Scale Vessels

Country	Number of Vessels	Percentage of Total FOC Flags	Cumulative Percentage
Spain	281	21	21
Taiwan	108	8	29
South Korea	88	6	35
Japan	81	6	41
United States	58	4	45
Spain (Canary Islands)	55	4	49
Norway	49	4	53
South Africa	45	3	56
Belize	42	3	59
Panama	42	3	62
All others	520	38	100
Total	**1,369**		

Box 3.3 Vessel Operations under Flags of Convenience

"Convenience countries . . . have neither the desire, the power, nor the administrative machinery to impose regulations that temper the narrowest private economic concerns with social objectives. If they tried to, they would cease to be convenient."[1]

Note

1. United States, House of Representatives, Armed Services Committee Hearing, Vessel Operations under Flags of Convenience and Their Implications on National Security. (2002). http://commdocs.house.gov/committees/security/has164220.000/has164220_0.HTM.

There are many other reasons why a fishing vessel would register under an FOC. Apart from the anonymity guaranteed to the beneficial owners of the fishing vessels, which allows them to fish illegally without much fear of legal repercussions, FOC-registered vessels do not have to pay for insurance, training of crew, safety gear, licenses, and vessel monitoring systems or observers,[24] among others. Therefore, using an "open registry" allows these fishing vessels to avoid regulations pertaining to health and safety,[25] insurance, and crew employment conditions,[26] as well as those pertaining to taxes, national and international legislation related to fisheries, and environmental and maritime laws and conventions.[27]

Illegal Fishers Use Ports of Convenience to Offload Fish Stolen from Their Waters

At some point, an IUU fishing vessel will have to arrive at port to offload or transship its catch, as well as to conduct other necessary functions, such as refueling and crew rotations. While this gives the opportunity for the port state to identify possible IUU fishing vessels, many such states do not have the necessary resources to properly monitor these activities. Port states refer to countries where fishing vessels enter to offload their catch. These states are responsible for regulating the offloading and transshipment of fish and enforcing relevant international regulations pertaining to IUU fishing should they detect any violations. They can specify conditions and limit or deny entry to port if they suspect that a vessel was engaged in such activities. One reason why a port may deny entry is if the vessel requesting access is a fishing vessel or a refrigerated cargo vessel that has been blacklisted by an RFMO (because of its history of IUU fishing). Most ports have established standardized measures that they use to inspect arriving vessels, much of which has been recommended by the UN Port State Measures Agreement discussed in more detail in Chapter 6. Among these are the requirements that the fishing vessel seeking entry into the port provide an advance notice, copies of the license to fish, details of their fishing trip, the catches on board, and so on.

Ports of convenience (POCs) collectively refer to ports with a free economic zone status.[28] As such, they generally have favorable customs regulations and lax controls over the transshipment of goods. POCs are attractive to IUU fishing vessels because they offer many customs advantages, which include exemptions from import duty, warehousing of goods with no time limits, and no customs procedures for goods leaving toward a third country. The lax controls over transshipment of goods at these ports make it especially difficult for both flag and port states to detect the activities of IUU fishing vessels.

In 2014, criminologists used the "risky facilities" approach to empirically test the characteristics of 121 POCs that were visited by illegal fishing vessels during 2004–2009. They tested the assumptions that those ports that facilitated "concealability" and "easy disposal of cargo" and that were in countries with "lax regulation and enforcement" attracted more IUU fishing vessels.[29] They found that the ports that had high levels of vessel traffic and those with larger harbor size (measures of "concealability"), and were located in countries with "lax regulation and enforcement," were the preferred ports for IUU fishing vessels to offload their catches. Table 3.3 shows the top 10 ports that IUU fishing vessels visited during 2004–2009 and the number of visits they made during this time (Figure 3.1 shows the locations of these top 10 POCs). In a follow-up study, the same researchers found that the offenders' "choice-structuring" decisions were affected by such "situational" factors as the quality of port infrastructure and the "concealability" provided by these ports. These vessels were, however, less likely to offload the catches in countries with strong governance.[30]

Table 3.3 Top 10 Ports of Convenience and IUU Vessel Visits to These Ports, 2004–2009

Country	Name of Port	Number of Visits	Percentage of Visits	Cumulative Percentage of Visits
Singapore	Singapore	32	7.5	7.5
Ecuador	Bahia de Manta	16	3.8	11.3
Ukraine	Sevastopol	16	3.8	15.1
Mauritania	Nouadhibou	15	3.5	18.6
Russia	Kaliningrad	15	3.5	22.1
Colombia	Cartagena	14	3.3	25.4
Germany	Rostock	14	3.3	28.7
South Korea	Pusan	12	2.8	31.5
Ghana	Tema	11	2.6	34.1
Spain	Las Palmas	10	2.4	36.5

Figure 3.1 The Geographic Distribution of the Top 10 Ports of Convenience

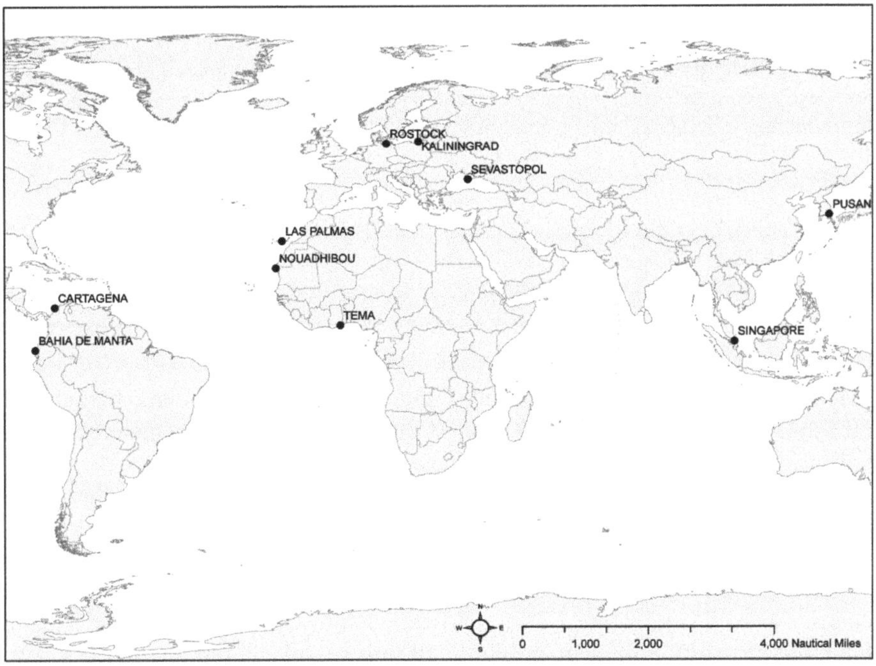

Particular controversy surrounds Las Palmas de Gran Canaria, a port located in Spain's Canary Islands, which serves as the largest point of entry for fish from West Africa coming into Europe. A study conducted by the Environmental Justice Foundation (EJF) found that 32 percent of the vessels entering the port (between January 2003 and December 2006) were linked to illegal fishing activities.[31] Once fish have been unloaded in Las Palmas, they can be transported anywhere within the European Union without further inspection of their origin or legality, thus allowing for an easy "laundering" of illegally caught fish into the legal market.[32]

Fish Stolen from Their Waters Are "Laundered at Sea"

Many distant-water fishing vessels stay on high seas for long periods of time. They can do so primarily because their catch can be transshipped through refrigerated cargo vessels, known as "reefers." Reefers are used to not only transport the catch but also refuel, rotate crews, and replenish bait, food, and water. Quite often, the same companies that operate the fishing vessels also own these reefers.[33] Illegal fishing vessels take advantage of this practice because it allows them to remain in distant waters for long periods of time.[34] Transshipping their catches at sea also allows these vessels to evade monitoring and enforcement at

Box 3.4 Transshipment at Sea

Transshipment at sea facilitates human trafficking, forced labor, and other human rights abuses because these vessels can remain in the open seas for extended periods of time and easily avoid detection and enforcement. Transshipment at sea has also been linked to trafficking in weapons, drugs, and other wildlife.[1]

Note

1. Ewell, C., Cullis-Suzuki, S., Ediger, M., Hocevar, J., Miller, D. & Jacquet, J. (2017). Potential ecological and social benefits of a moratorium on transshipment on the high seas. *Marine Policy, 81*: 293–300.

sea. In West Africa, field investigations undertaken by the EJF found that almost all illegal reefers were flying FOCs. Of the total 700 reefers registered under FOCs, 70 percent carried the flags of Panama, Bahamas, and Liberia.[35]

When an IUU fishing vessel transships its catches at sea and uses these vessels to offload a mix of legally and illegally caught fish, it becomes virtually impossible to distinguish between these fish when the reefer is inspected at port. Transshipping illegal catches at sea, therefore, allows these vessels to evade port controls as well (see Figure 3.2 for a schema on how fish laundering can happen via transshipping). Similarly, an illegal fishing vessel can use a reefer that also carries fish caught legally by other vessels. In fact, a recent study that tracked global transshipping behavior during 2012–2016 found that a significant number of transshipping activities happened around regions that are also known to have a high degree of IUU fishing. For example, the study recorded 11,000 transshipping events in East Central Atlantic, which is the area that has the highest IUU fishing activity (37 percent). Importantly, a majority of these transshipping

Figure 3.2 How Does Transshipment at Sea Help Launder IUU Fish?

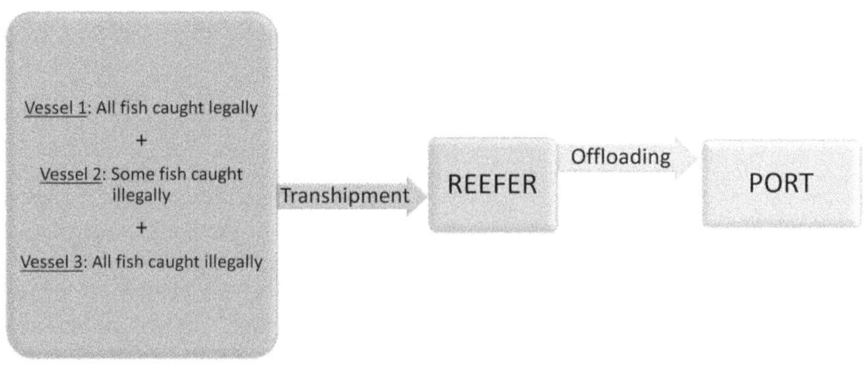

Box 3.5 Container Ships

A recent study[1] examining the situation of IUU fishing in West Africa has uncovered a new trend. Instead of using refrigerated cargo vessels to transport (likely illegally caught) fish from West Africa into the European Union, 84 percent of the fishers who caught fish from West African waters used *refrigerated container ships* to move the fish into the European Union. They used major container hubs that included Walvis Bay (Namibia), Cape Town (South Africa), Dakar (Senegal), Abidjan (Côte d'Ivoire), Lagos (Nigeria), and Tema (Ghana) to move the fish from the fishing vessel into the container ship. This makes it virtually impossible even for the well-equipped EU ports to detect illegally caught fish because the EU IUU Regulation that imposes the restrictions on the use of refrigerated cargo vessels by its Member States does not extend these restrictions on container vessels. Container vessels are exempt, and this severely compromises the inspectors' ability to detect possible violations.

Note

1. Daniels, A., Gutiérrez, M., Fanjul, G., Guereña, A., Matheson, I. & Watkins, K. (2016). *Western Africa's Missing Fish: The Impacts of Illegal, Unreported and Unregulated Fishing and Under-Reporting Catches by Foreign Fleets*. London, UK: Overseas Development Institute.

activities (57 percent) occurred within the countries' exclusive economic zones (EEZs), and these reefers tended to visit the ports near the location of the transshipping activity. About 40 percent of transshipping events involved vessels flying FOCs.[36] Considering no foreign power will be allowed to board, seize, or arrest the crewmembers of these vessels[37] if caught transshipping illegally, the offloading of illegal cargo from these vessels, therefore, can go largely unpunished.

Realizing the complexity of the traceability of fish transported using such vessels, many countries have banned this practice within their EEZs, such as Senegal and Côte d'Ivoire in West Africa[38] and Thailand and Indonesia in Southeast Asia.[39] The Forum Fisheries Agency also banned transshipping at sea (particularly those involving purse seine vessels) in the Pacific Islands, mandating that such activities only take place at port.[40] The European Union restricted this practice for Member State flagged vessels. The European Union allows transshipping at sea when the vessel's activity is closely monitored by an RFMO through observers onboard the vessel.[41] Certain RFMOs, such as the Western and Central Pacific Fisheries Commission, have introduced high seas boarding schemes in their regulatory provisions to deal with the problem of transshipment in the high seas.[42] Another RFMO, the South East Atlantic Fisheries Organization, introduced a complete ban on transshipment in 2006.[43] However, transshipment is poorly monitored in many parts of the world, and many scholars agree that unless a moratorium on transshipment is introduced, not much can be done to effectively deal with this global problem.[44]

They Lack Essential Monitoring, Control, and Surveillance (MCS) Capacity to Protect Their Waters

This is the view of major nongovernmental organizations, such as Greenpeace, World Wildlife Fund, and TRAFFIC, and RFMOs that deal with the problem of IUU fishing on a daily basis. This view was expressed in a report[45] prepared by the government of Australia and the UN Food and Agriculture Organization that surveyed these organizations with the attempt to understand key drivers that contribute to the problem. These organizations showed concern that developing countries often lacked adequate financial resources to effectively regulate and monitor their EEZs. This situation is further exacerbated by the fact that FOC countries are unwilling to cooperate with these developing countries to address the problem of IUU fishing carried out by the vessels carrying their flag.

A similar story emerges when examining 49 countries in the North Pacific, Southeast Asia, Western and Central Pacific, and South Eastern Pacific regions. Ineffective MCS, as well as ineffective domestic management capacity, coupled with the lack of alternative employment opportunities, political will, and sound national legislation, has hindered the efforts in these countries to effectively deal with the IUU fishing problem within their jurisdictions.[46] The 13 countries of Southeast Asia are also in a constant battle over un-delimited boundaries and have to deal with broader political sensitivity over some regional issues, complicating efforts to effectively deal with IUU fishing. Their lack of shared vision across adjacent states to improve fisheries management and more effectively deal with IUU fishing have served as significant obstacles to combat IUU fishing. Meanwhile, the 16 economies in the Western and Central Pacific region suffer from more unique management problems that pertain to high seas fleets, flag state controls, and specific access arrangements. These countries also lack effective resources to coordinate and communicate across national agencies.

It Is Economically Beneficial to Engage in Illegal Fishing in Their Waters

Researchers have devised an economic model[47] to understand and explain the costs and benefits of engaging in IUU fishing. They examined such factors as the (a) benefits of engaging in the illegal activity, (b) probability of being caught, (c) penalty if caught, (d) cost the fisher would pay to avoid detection, and (e) degree of the fishers' moral and social standing in society. Their model suggests that fishers have about 20 percent chance of being caught; however, if they do get caught at that frequency, it is still financially beneficial to engage in IUU fishing. The fines in place hardly deter IUU fishers—these fines will have to be increased by 24 times for the expected costs to be at least as much as the expected benefits, leaving IUU fishers with little incentive not to engage in such activities.

They Lack Scientific Basis When Making Fisheries Management Decisions

This is the view expressed in a survey conducted with 1,188 fisheries managers, academics, and governmental and nongovernmental organizations from every coastal country in the world.[48] The survey attempted to understand the importance of sound fisheries management practices in dealing with not only IUU fishing but also fisheries sustainability in general. The researchers found that of all the fisheries management practices worldwide, only a few have been effective in their management efforts, and these fisheries shared the following characteristics in common: they (a) used robust scientific basis for management recommendations, (b) converted the scientific recommendations into policy and practice, (c) ensured compliance with regulations in place via robust enforcement, and (d) minimized the extent of subsidies and fishing overcapacity and, importantly, foreign fishing by issuing a limited number of fisheries access agreements.

Nearby Countries' Practices Matter

While examining the compliance of 53 countries with the UN Code of Conduct for Responsible Fisheries, researchers not only examined the factors that had already emerged as significant drivers of IUU fishing in many prior studies (namely corruption, a country's GDP, efforts made to protect biodiversity, and fish stocks export share within the global fishing industry) but they also studied the unique spatial dependency of countries and whether this factor relates to IUU fishing. Interestingly, these researchers found that a country is more likely to comply with international fisheries regulations in dealing with IUU fishing if it is in close proximity to countries that are willing to comply as well. The country is also more likely to partake in local and regional enforcement agencies if nearby countries do the same.[49] This leads to a significant and important conclusion: any international measures aimed at fisheries compliance on the global scale should note the role the local and regional institutional arrangements play in imposing fisheries regulations.

Illegal Fishing Is Significantly Higher in Countries That Have to Deal with Corruption and Other Governance Issues

Many researchers have expressed concern about how corruption facilitates overexploitation of marine resources and significantly hampers the efforts to effectively enforce existing fisheries regulations. This is especially true in developing countries because in these countries, corruption is widespread and government enforcement is weak.[50]

Africa is one place where corruption is rampant. The sectors most vulnerable to corruption are those dealing with licensing and access agreements, and the most common types of corruption include bribery, embezzlement of license fees by public officials, conflict of interest, and undue influence.[51] In these sectors, fisheries officials, as well as other high-level officials, exercise discretionary power when

Box 3.6 Bribery

"I understand why people don't trust the inspectors. It's easy to get away with everything. First they make you feel afraid, saying that you will lose your license. Then they say that something can be done. So you pay them. That's why people know that on a bad day it's ok to take some of the crays under the minimum-size."[1]

Note

1. Sundstrom, A. (2013). Corruption in the commons: Why bribery hampers enforcement of environmental regulations in South African fisheries. *International Journal of the Commons, 7*(2): 454–472.

dealing with fishing vessels, facilitating IUU fishing carried out by these vessels. Similarly, in the Pacific Islands, corruption in the fisheries sector is pervasive. The sectors where corruption is most prevalent include licensing, access agreements, and monitoring and inspection. The following are some case study examples used to illustrate how corruption facilitates IUU fishing in these regions.

Case Study 1. Observers are fisheries managers who are assigned to accompany foreign fishing vessels to ensure that fishing rules set forth in the access agreement between their country and that of the foreign fishing vessel are being followed. An investigation conducted by the EJF and Al Jazeera revealed that, in Sierra Leone, in many cases, observers' salaries are paid by the fishing companies whose activities the observers are actually assigned to monitor.[52]

Case Study 2. In South Africa, a Korean captain fishing illegally attempted to bribe an observer. The captain was brought to court, but later released after paying an insignificant fine.[53] Speculations suggested that this captain had bribed the prosecutor and the judge in the South African court.

Case Study 3. Fish operators in South Africa received bribes to falsify licenses and documentation that allowed more than 600 tonnes of illegally caught rock lobster and approximately 2,800 kg of Patagonian toothfish to be exported to the United States.[54]

Case Study 4. In 2005, in Solomon Islands, the Minister of Fisheries and Marine Resources declared in Parliament that the government dismissed a number of permanent secretaries following the revelation that they had been engaged in diverting license fees into personal overseas private bank accounts, and these fees were received from overseas fishing companies in return for favorable license conditions.[55]

Case Study 5. Pacific Island senior officials received handsome payments after settling access negotiations with foreign countries. Their payments included first-class plane tickets and lavish hotel accommodations and entertainment in the foreign country, extended holiday and other trips, as well as payment of overseas tuition fees for their children.[56]

Case Study 6. In Pacific Islands, port inspectors were offered a large tuna in return for "going easy" on the vessel and not examining the fishing vessel's logbooks and turning a blind eye to infractions with licensing conditions.[57]

Case Study 7. In 2011, a South Korean illegal fishing vessel fishing in Liberia was released following pressure from the Korean embassy. Liberia released the vessel and the illegal catches due to diplomatic pressure.

Case Study 8. In 2004, a Tanzanian patrol vessel discovered 25 EU-flagged vessels operating illegally in the country's protected areas. Following the arrest of the crews of these vessels, a delegation from the European Commission's Directorate-General for Maritime Affairs and Fisheries visited Tanzania, after which the vessel's crews received a fine, and the amount of that fine remains publicly unavailable.[58]

Case Study 9. In Angola, oftentimes, legal fishers and citizens do not report illegal fishing activities because they know that the vessel engaged in these activities is tied to public officials or operated by employees of the Ministry of Fisheries.[59]

Case Study 10. In one East African country, it is widely known that the three leading commercial fishing companies catching in-shore prawn are co-owned by the president and the current and former Minister of Fisheries.[60]

Case Study 11. In South Africa, an investigation carried out by the Scorpions—a specialized government-organized crime unit—revealed that Hout Bay Fishing—a major multinational fishing company operating in the country—was buying illegally caught fish, as well as systematically fishing over its quota during the 1999–2000 fishing season. The company sold Patagonian toothfish, lobster, and hake worth millions of dollars. This was possible because the company relied heavily on 14 marine conservation inspectors who received bribes in return for falsifying catch and export data.[61]

Case Study 12. In 2007, Sierra Leonean navy arrested eight individuals from Conakry, Guinea who were accused of engaging in a "pirate" attack on two Chinese trawlers that were licensed by the country to fish in its waters. Military, police, and fisheries officers, armed with AK-47s, were among those arrested in this attempted "pirate" attack.[62]

Case Study 13. In 2009, there were widespread allegations of corruption in Senegal that involved about $9 million paid as a result of licensing arrangements with Russia. President Abdoulaye Wade was believed to have embezzled the money to finance elections and buy off votes. His son, who held multiple ministerial positions, was later arrested and faced a highly publicized trial over the unexplained $240 million wealth.[63]

Case Study 14. In 2009, a district treasurer in the African National Congress (ANC) was stopped in his car by the police. In the back of the car, the police found about 2,500 shucked abalone worth about $28,000, and the bags were plastered with ANC branding.[64]

These case studies clearly demonstrate the pervasiveness of corruption in developing countries that significantly impacts the fisheries, hampers enforcement efforts, and puts fishers and coastal communities' livelihoods at stake. The lack of transparency and the secretive nature of access agreements, the one-man role assigned to run a system of licensing and the inability, and often the unwillingness of the developing countries to properly manage their resources create ripe conditions for those in power to abuse their authority and corrode the intentions of those who are willing to comply with fisheries regulations. Importantly, all efforts to deal with IUU fishing are undermined by several actors who have the power to actually do something to effectively deal with the problem.

WHAT ARE SOME OF THE SPECIFIC REASONS WHY ILLEGAL FISHING OCCURS IN PARTICULAR REGIONS OF THE WORLD?

Apart from the factors facilitating IUU fishing globally, there are some important regional patterns that have emerged as additional reasons why IUU fishing occurs in these regions. The factors discussed in the text are unique to the regions from which the case studies are drawn and may not be applicable globally or to other regions of the world.

Case Study 1: EU States Passed National Laws to Circumvent EU Regulations

In 2007, the EJF[65] ran an investigation to uncover the problem of illegal drift net fishing in the Mediterranean Sea. Through this investigation, the organization discovered a worrying detail. With a closer look at Italy, France, Morocco, and Turkey, the EJF discovered that Italy and France have deliberately passed laws to circumvent EU and international drift net bans, thus allowing illegal activity in the Mediterranean Sea to continue without significant legal consequences. A major driver for this was, undoubtedly, the increasing demand for swordfish, caught in the Mediterranean Sea, and this demand was met by not only Italy and France but also other Mediterranean countries, including Morocco, Turkey, Spain, Malta, Greece, and Algeria.

Case Study 2: IUU Fishing Is Culturally Tolerated in the Peninsular Malaysian States

In 2008, Asia-Pacific Economic Cooperation (APEC) studied the prevalence of IUU fishing in the East Coast region of Peninsular Malaysian states, namely Kelantan, Terengganu, Pahang, and eastern Johor, and the factors driving such activities in the region.[66] APEC discovered not only that IUU fishing of protected species, such as lobsters, arowana, cockle spat, turtle eggs, grouper fry, and sharks, was prevalent both in national waters and, alarmingly, in marine-protected areas but also that this activity was culturally acceptable. There was generally a cultural tolerance of "rule bending," and the significant reasons why

IUU fishing was acceptable in the region were cultural cuisine habits and, importantly, beliefs in medicinal properties from marine organism, leading to turning a blind eye to the alarming rates of the plunder of fish from their waters.

Case Study 3: Having Easy Access to Dynamite and Cyanide Helps Fish Illegally in the Sulawesi Sea

The Sulawesi Sea is one of the most significant and biologically diverse marine areas in the Asia-Pacific region. The general factors of poor monitoring and surveillance in the region, high demand (both in the domestic and foreign markets) for aquarium and exotic fish that can be found in the Sulawesi Sea, and social conditions among the coastal communities in the region are all important contributing factors to IUU fishing. What is most alarming, however, is the availability and access to substances that can be used to make dynamite, such as ammonium nitrate and blasting caps, and poisons, such as cyanide. These have made IUU fishing in the Sulawesi Sea relatively easy. Coupled with the lack of maritime boundary agreements among APEC states, using easily accessible illegal means of removing fish has exacerbated the control and surveillance efforts in the region.[67]

Case Study 4: Pacific Island States Are Unable to Compete with Powerful Countries to Protect Their Economic Interests

The tuna fisheries in the Pacific Islands region and the tuna fisheries in the central Pacific Ocean together comprise the most important tuna fisheries in the world.[68] However, the Pacific Islands region is incapable of fighting the IUU fishing of tuna, and the primary reason for this is the inability of the states within the region to compete with powerful countries to protect their economic interests. In addition to this, poor national fisheries governance, corruption, lack of skilled fisheries managers, poor policies related to the management of tuna resources, a plethora of issues related to market access, and lack of regional solidarity among the Pacific Island states have all created ripe conditions for tuna plunder by not only nationals but also foreign fishing fleets..

WHAT ARE IUU FISHERS SAYING?

Case Study 1: It Pays to Fish Illegally, and There Is No Moral Obligation to Do Otherwise

In 1998, researchers interviewed 318 fishers in Peninsular Malaysia with the goal of understanding the underlying reasons why these fishers were likely to violate various fisheries regulations.[69] These researchers discovered that fishers made economic calculations of the actual likelihood of being caught and the applicable penalties if caught. These fishers suggested that, in fact, it was much more profitable to violate the law than to comply to it. Importantly, the fishers and the communities where they lived placed little legitimacy on the fisheries regulations in place and these regulations' actual regulatory power. In other words, these regulations meant little to them and hardly kept them back from fishing illegally. Apart from this important factor, witnessing other fishers' chronic violations of fishing regulations made it morally and socially acceptable to engage in similar illegal behavior.

In 2003, researchers made a similar discovery when interviewing Danish fishers involved in the cod fishery in the Baltic Sea, the lobster fishery in Kattegat, and the industrial fishery in the North Sea.[70] Through a study of 154 fishers in these regions, these researchers discovered that they were likely to engage in illegal fishing because of the obvious economic gains and low likelihood and perceived inefficacy of sanctions in place. These fishers showed no reservations pertaining to noncompliance behavior and referred to the illegal behaviors of other fishers, which were morally acceptable, as other likely causes of their illegal behavior. They also suggested that the lack of compatibility between the regulations and actual fishing practices and patterns was the main reason why they engaged in illegal fishing. Their noncompliance was also, in a way, a rebellion against the regulations that were put in place without making these fishers part of the decision-making process of what would be a more appropriate way to regulate these types of fisheries.

Lastly, a study[71] conducted in the United States in 2009 attempted to understand the dynamics of compliance, as well as enforcement challenges, failures, and successes in four fisheries that included the Northeast multispecies groundfish fishery, the Gulf of Mexico red snapper fishery, the Pacific groundfish trawl fishery, and the Bering Sea/Aleutian Islands pollock fishery. The researchers found that up to one-third of fishers routinely violated fisheries regulations, while another one-third of fishers were occasional violators. In other words over 60 percent of fishers violated fisheries laws at least once. Even in this developed country known for its high regulatory and enforcement capacity, the majority of fishers were less likely to comply with fisheries regulations because, according to them, there was (a) low probability of a violation being detected and actually prosecuted, (b) significant incentive to not comply because the expected costs of penalties were significantly lower than the economic benefit of engaging in illegal fishing, (c) lack of moral obligation and social pressure to comply, (d) notable failure at at-sea enforcement to deter many types of violations, (e) notable failure to prosecute chronic and substantial violators, and (f) unavailability of adequate enforcement data and communication between fisheries managers and law enforcement personnel, among others. These reasons made it easier to not only morally justify their illegal activities but also persist in such activities without much fear of getting caught.

Case Study 2: Indigenous People as Illegal Fishers?

In 2010, a study conducted in Raja Ampat Regency, eastern Indonesia, discovered an alarming pattern.[72] The laws governing fishing of the resources in the regions intended for subsistence and traditional fishing vessels were ambiguous, and these laws were hardly enforced. As a result, large local and foreign fishing vessels increasingly moved into these fishing grounds, intended only for subsistence and traditional fishing, to carry out their illegal activities. Small-scale vessels followed. Eventually, the indigenous people in the region, who previously engaged in subsistence fishing allowable by law, became increasingly "integrated into the cash economy"[73] and began illegally exploiting the fish stock resources for commercial gain.

CONCLUDING REMARKS

As with any illegal activity, IUU fishing is likely to happen when there are conditions facilitating it. These facilitating conditions not only encourage these fishers to engage in the illegal activity but also make it easy for the fishers to persist in the illegal behavior with little fear. Illegal fishing is made possible through such widely used practices as operating under an FOC, which allows IUU vessel owners to avoid national and international fisheries laws; using a POC, that is, a port with lax customs regulations, to unload illegal catch; or transshipping the illegal catch through the use of refrigerated cargo vessels.

Illegal fishers are also aware of other existing loopholes that help them operate virtually undetected and with no fear of getting caught: they target developing countries where corruption is rampant and there is lack of political will to enforce fisheries regulations; they know to calculate the countries' lack of MCS capacity against the probability of getting caught if fishing illegally in these countries' waters; they are largely aware of the cultural tolerance of "rule bending" if they are to fish illegally locally; they are aware of much of the illegal gear and substances that are largely available to them; and they are essentially driven by the significant economic incentives and the growing demand for wild-caught fish. These illegal fishers are simply calculating, opportunity-driven "entrepreneurs" ready to exploit fishing grounds, especially when it is remarkably easy to do so. The combination of increasing global demand for fish and the availability of the means to avoid both national and international fisheries laws, therefore, creates ideal conditions for these fishers to engage in illegal fishing globally.

Chapter 4

The Likely Causes of Illegal Fishing

APPLYING PRINCIPLES OF ENVIRONMENTAL CRIMINOLOGY

Environmental criminology refers to a group of theories that argue that crime events should be understood in terms of environmental influences and that crime patterns should be analyzed through the "sociodemographic, temporal, and spatial qualities"[1] of these environments and other immediate circumstances that are likely to make the commission of crime possible. In other words, environmental criminologists suggest that one should focus on the *crime events* and how and why these events unfold rather than on individuals' *criminality* or *criminal motivations*.

The Rational Choice Perspective

One of the theories within the family of environmental criminology is the rational choice perspective proposed by criminologists Derek Cornish and Ronald Clarke.[2] They argue that when thinking about crime, one should focus on the decision-making process of likely offenders, which are highly specific, and they break down this process into two phases (shown in Table 4.1).

In Phase 1 of crime involvement, such factors as low self-control, weak social bonds, and previous learning experiences play an important role. These factors create the inclinations and criminal motivations to engage in crime. This phase is further divided into (a) initial involvement (initiation), (b) continued involvement (habituation), and (c) desistance stages. At each of these stages, as offenders make decisions, different sets of factors affect their decision. While, for example, in the initiation stage a person's individual background may play a significant role (e.g., upbringing and motivation to commit a crime), the habituation stage is primarily affected by the offender's success in committing the initial criminal act and desistance is likely to be determined by the failure of the initial crime.

In Phase 2, the environmental cues and specific *situations* within the immediate environment are the key drivers. These environmental cues signal the *opportunities* for crime, as well as trigger the calculations about whether it is

Table 4.1 The Two Phases of the Decision-Making Process to Engage in Crime and Factors Influencing It

Phases	Factors Influencing the Decision Process	How Do These Factors Influence the Decision Process?
Phase 1: Initial involvement	Low self-control, weak social bonds, class origin, intelligence, previous learning experiences, family upbringing, and neighborhood background	These create "criminal motivations, deeply-rooted inclinations or dispositions to commit crime."[1]
Phase 2: Criminal event	Immediate environmental cues and specific situations	These trigger "crime-specific" calculations of cost and benefit of engaging in crime.

[1]Clarke, R. V. & Cornish, D. B. (2001). Rational choice. In Paternoster, R. & Bachman, R. (Eds.). *Explaining Criminals and Crime*. Los Angeles, CA: Roxbury Publishing Company, p. 33.

worth engaging in crime. Therefore, it is the *opportunities* presented to the offenders during the normal patterns of social and economic life that play a significant role in their decision-making process.[3] A decision *not* to commit a crime is, therefore, also made when the calculated risks are higher than the anticipated rewards.

The rational choice perspective warns that one should never treat a criminal act as senseless or irrational: a criminal act is purposive and deliberate, committed to benefit the offender, and offender decision-making is always different for different types of crimes. Although these *opportunities* and situations may be constrained in time and space, these dynamics are factored into the decision-making formula, and it is these factors that may consequently constrain the potential offender's cognitive abilities.[4]

Of course, offenders cannot calculate the costs and benefits of committing a crime with precision. They do so with "limited rationality"[5] because they cannot consider *all* the possible costs and benefits when deciding to engage in a crime.

Box 4.1 Rational Choice

"In seeking to benefit themselves, offenders do not always succeed in making the best decision because of the risk and uncertainty involved."[1]

Note

1. Clarke, R. V. & Cornish, D. B. (2001). Rational choice. In Paternoster, R. & Bachman, R. (Eds.). *Explaining Criminals and Crime*. Los Angeles, CA: Roxbury Publishing Company, p. 24.

Any given crime involves an array of benefits (e.g., financial reward, prestige, status, revenge, excitement, and sexual gratification) and costs (e.g., financial loss, arrest, and loss of reputation). Therefore, offenders' calculations are limited to their moods, feelings, immediate motives, and intentions, as well as the amount of time available to them to engage in a criminal behavior.

The rational choice perspective also applies to the organizational and corporate decision-making processes. Like individuals, organizations behave based on the risk-reward and cost-benefit calculations, which are considered in the background of the following four guiding questions[6]:

1. What possible actions can they engage in (thus alternatives)?
2. What possible consequences can follow based on these alternatives (thus expectations)?
3. Which alternative is more preferable based on how valuable it may be (thus preference)?
4. What would the decision-making process entail in terms of choosing the possible alternatives (thus the decision rule)?

Like individuals, organizations are also conscious decision makers who weigh their options and act within the rationality of their calculations, however limited this calculation may be. It is within this context that a crime is more likely to occur.

The major arguments of the rational choice perspective can be summarized as follows[7]:

- Criminal behavior is purposive.
- Criminal behavior is rational.
- Criminal decision-making is crime specific.
- Criminal choices fall into two broad groups: "involvement" and "event" decisions.
- There are separate stages of involvement.
- Criminal events unfold in a sequence of stages and decisions.

A purposive behavior refers to any behavior that is committed with the intention of benefiting the offender. A rational behavior means individuals are likely to select the best available means to achieve their motives and goals. These motives and goals are very crime specific and are determined by the crime being committed. For example, rape may be committed with the purpose of gratifying the desire for control and dominance, while theft satisfies the need for cash or goods. Crime "involvement" decisions pertain to the offender's criminal career and include decisions made during the initial involvement (i.e., initiation), continued involvement (i.e., habituation), and desistance stages of crime commission. Lastly, crime "event" decisions are crime specific and include a series of choices that are made when preparing for and carrying out a specific criminal act.

Situational Crime Prevention

The situational crime prevention (SCP) perspective suggests that when thinking about crime, one should focus on the particular *situations* and *opportunities* that are likely to make the commission of crime easy.[8] SCP argues that to understand why crime occurs, there is no need to pay attention to criminal motivations because this "dispositional bias" offers little in terms of the preventive measures and practical solutions that can be adopted to address crime. This perspective also notes that crime opportunities are not distributed randomly but rather concentrate in time and space: only certain locations are more crime prone than others, and crime occurs during specific time periods. Therefore, specific situations and environments lead to specific opportunities for the commission of specific crimes.

The perspective makes the following several important claims:

- Opportunity plays a part in every form of crime.
- Opportunity is an important cause of crime.
- Criminally disposed individuals will commit a greater number of crimes if they encounter more criminal opportunities.
- Regularly encountering such opportunities could lead these individuals to seek even more opportunities.
- Individuals without preexisting dispositions can be drawn into criminal behavior by a proliferation of criminal opportunities.
- Generally law-abiding individuals can be drawn into committing specific forms of crime if they regularly encounter easy opportunities for these crimes.
- The more opportunities for crime that exist, the more crime there will be.
- Reducing opportunities for specific forms of crime will reduce the overall amount of crime.[9]

Consequently, if it is the opportunity that leads to crime, then crime prevention strategies should be opportunity reduction techniques

1. that are directed at highly specific forms of crime.
2. that involve the management, design, or manipulation of the immediate environment in as systematic and permanent a way as possible.
3. so as to reduce the opportunities for crime and increase its risks as perceived by a wide range of offenders.[10]

Situational crime prevention can be achieved via two general categories of situational factors.[11] The first category involves reducing the physical opportunities needed to commit crime, for example, putting steering column locks on vehicles to make theft difficult. The second category involves increasing the risks of being caught, for example, "raising the chance of an offender being seen by someone who is likely to take action."[12]

SCP offers strategies that can be used to block crime occurrence. These strategies, captured in 25 crime prevention techniques, focus on (a) *increasing the perceived effort* needed to commit the crime, (b) *increasing the perceived risk* of committing a crime, (c) *reducing the anticipated rewards*, (d) *reducing provocations*, and (e) *removing excuses*.

Increasing the Perceived Effort

Most crimes, especially property crimes, are easy to commit and do not require a complex set of skills.[13] Therefore, deflecting offenders into noncriminal actions is easy through a set of techniques that focus on (1) hardening targets, (2) controlling access to facilities, (3) screening exits, (4) deflecting offenders, and (5) controlling tools or weapons. Examples of target hardening to reduce traditional crime include using tamperproof packaging on items that are stolen most frequently. Controlling access involves measures designed to exclude potential offenders from places, for example, using personal identification numbers to access a bank account. Screening exits may include requiring export documents, electronic merchandise tags, and tickets when exiting a particular area. Deflecting offenders can be achieved by placing street closures to prevent access to restricted areas. Lastly, restricting sales of guns to individuals under a certain age is a way to control tools.[14]

Increasing the Perceived Risk

Increasing the perceived risk can be achieved by extending guardianship of places or people in a way that will discourage an offender from committing a crime for the fear of being seen or caught. This "deterrent threat"[15] can be achieved through (6) extending guardianship, (7) assisting natural surveillance, (8) reducing anonymity, (9) using place managers, and (10) strengthening formal surveillance. One can extend guardianship by going out in groups at night. Enhancing street lighting or lighting inside public housing are examples of natural surveillance, so are designing areas that increase visibility by trimming bushes. Anonymity can be reduced, for example, by requiring taxi drivers to display personal identification information and license documents. Place managers, that is, individuals who are responsible for a site, can increase the perceived risk by closely monitoring the workplace. These may include convenience store clerks, parking lot attendants, residential or commercial building doorkeepers, and train conductors. Lastly, examples of formal surveillance measures include using closed circuit televisions (security cameras), burglar alarm systems, police, security guards, or store detectives assigned to high-crime areas.

Reducing the Anticipated Rewards

Crime commission is less attractive when the motivated offender does not know of the expected reward. If, for example, a potential offender does not know what is contained in a sealed, unmarked box in a store, he or she is less likely to steal it. Similarly, items that are marked are less likely to be stolen. Therefore, anticipated rewards can be reduced by (11) concealing targets, (12) removing targets, (13) identifying property, (14) disrupting markets, and (15) denying benefits. Targets can be concealed or removed by placing frequently stolen items, such as razor blades or cigarettes, behind the counter. Marking an electronic device, such as an iPhone with a personal message, is an example of identifying property. Disrupting markets can be achieved by closely monitoring outlets, such as pawnshops, where stolen goods may end up. Lastly, denying benefits could involve quickly painting out graffiti.

Reducing Provocations

This technique emphasizes the importance of understanding the immediate triggers for criminal events.[16] Reducing these immediate triggers is likely to prevent crime from happening through (16) reducing frustration and stress, (17) avoiding disputes, (18) reducing temptation and arousal, (19) neutralizing peer pressure, and (20) discouraging imitation. Forcing people into efficiently monitored lines can reduce frustration and stress. Disputes can be avoided when bars make the effort to reduce crowding. Temptations can be reduced if frequently stolen items are placed behind glass doors and can only be accessed with the help of a store clerk. Neutralizing peer pressure could involve dispersing bullies at school or setting specific standards of behavior to reduce bullying. Lastly, quickly repairing "broken windows"[17] or removing signs of vandalism will discourage imitation.

Removing Excuses

This strategy is designed to make it difficult for someone to use the "I did not know" excuse for criminal behavior. It is also designed to make complying with laws and regulations easy and with least effort.[18] This strategy's five techniques include (21) setting rules, (22) posting instructions, (23) alerting conscience, (24) assisting compliance, and (25) controlling drugs and alcohol. Examples of setting rules include regulating employees' access to telephones or Internet sites. Such signs as "Private Property" are examples of posting instructions. Requiring signatures for customs declarations and posting signs such as "shoplifting is stealing" and "these premises are protected by video surveillance" are examples of alerting conscience. Placing litter receptacles in public places and using signs or physical barriers to control crowds are examples of assisting compliance. Drugs and alcohol can be controlled by using breathalyzers in bars.[19]

HOW CAN THE RATIONAL CHOICE PERSPECTIVE HELP EXPLAIN ILLEGAL FISHING?

If, indeed, illegal fishers are rational and make cost-benefit calculations, what would these calculations possibly involve? First, we will assume that the potential offender will consider the *degree of effort* involved in getting the caught species to the intended markets. If the target species are within the exclusive economic zone (EEZ) of a country with strong port inspection programs or in a country that is too far from a port of convenience, this will discourage the potential illegal fisher from engaging in illegal fishing. Second, the offender will calculate the *possible reward* by considering the availability of the highly sought-after resource (e.g., commercially significant fish that can be sold fast and for a reasonably high price), regardless of the effort. Third, the offender will calculate the *possible risk* of being caught. This risk calculation will entail not only the risk of being caught while fishing illegally at sea but also the risk of being detected with an illegally caught fish onboard the vessel while offloading at port. Although a country may have rich resources of commercially significant fish, it may not be vulnerable to illegal fishing if it has sufficient surveillance measures in place both at sea and at landing ports. So, if we look at illegal fishing through the lens of the rational choice perspective, we would expect to see that illegal fishers will not select the locations to engage in illegal fishing at random but rather go where there is a significantly lower chance of being caught (both at sea and at ports) and significantly higher reward in terms of the availability of the fish sought.

HOW CAN SITUATIONAL CRIME PREVENTION HELP DEAL WITH ILLEGAL FISHING?

Situational crime prevention techniques that have been widely used to deal with traditional crime problems can also be applied to illegal fishing. Many of these tactics are discussed in Chapter 6 with more specific recommendations. Here, we will briefly mention some. The technique "controlling access to facilities" can be achieved by requiring the installation of vessel monitoring systems and by inspecting vessel licenses. "Screening exits" would entail inspecting catches at ports before the vessels (both fishing vessels and reefers) are allowed to offload their catch. The "reducing anonymity" technique applies to requiring vessel registration. "Strengthening formal surveillance" would incorporate conducting inspections at sea by patrol vessels or observing catches via aerial surveillance. Meanwhile, "utilizing place managers" would mean using observers on board while fishing, who would monitor the vessels' activities by keeping logbooks on catches, recording fishing gear used, and making notes on other activities to identify potential violations. In terms of preventing illegal fishing behavior both within countries' EEZs and in the high seas, situational crime prevention offers many helpful techniques.

CONCLUDING REMARKS

One of the most difficult questions to ever answer is the "why" question. It is much easier to explain the what's, where's, and how's of crime, as these require more concrete and relatively easily derivable answers. The "why" of crime is much harder to understand, as these "why"s can be multiple, complex, and varied. Theory often helps channel the "why"s into more clear explanations. The two perspectives discussed in this chapter, the rational choice and situational crime prevention, are one such channel that assist in thinking about why one would engage in illegal fishing. Thus, relying on these perspectives, we propose that illegal fishing will be concentrated in places where there is a greater opportunity to do so. If this is, indeed, true, then we would expect to see that

- illegal fishing will concentrate in places that are abundant in highly commercial fish.
- illegal fishing will concentrate in places where less effort is necessary to get to these highly commercial fish.
- illegal fishing activities will continue as long as incentives to engage in it persist. These incentives are the expectations that the expected benefits will exceed the expected costs.
- offenders' decision will not only be guided by the calculation of the availability of highly commercial fish but also be weighed against the likelihood of being detected.
- illegal fishing is less likely to occur within countries with effective observer programs, port inspection strategies, and other access-control measures.
- effective surveillance both at sea and at ports is an important mechanism that would factor into the decision-making process.
- countries that are within a close geographic proximity to a number of ports of convenience will be preferred over other countries by international illegal fishing vessels, given they have the same resources.

Chapter 5 will, thus, test these theoretically driven assumptions.

Chapter 5

What Increases or Reduces the Risk of Illegal Fishing?

To better understand the global trends of illegal fishing and the possible reasons behind these trends, this chapter conducts in-depth analyses of data on illegal fishing activities in the exclusive economic zones[1] (EEZs) of 53[2] countries. These countries were selected because 96 percent of the global fish catches occur within their EEZs.

There are a total of 150 coastal countries in the world, which are sovereigns of 232 EEZs. In addition to their mainland, several countries are sovereigns of overseas territories; therefore, they have multiple EEZs. For example, in addition to its mainland, Australia holds sovereignty over several islands that include Heard Island and McDonald Islands, Christmas Island, Cocos Islands, Norfolk Island, and Macquarie Island. In this chapter, the analyses will relate to the activities that take place within the countries' mainland EEZs and not their overseas territories.

THE ROLE OF STRONG MONITORING, CONTROL, AND SURVEILLANCE CAPACITY IN REDUCING THE RISK OF ILLEGAL FISHING

What Is Monitoring, Control, and Surveillance in the Context of Fisheries Management?

Originally, monitoring, control, and surveillance (MCS) specifically referred to measures of law enforcement and applied to actions involving boarding and inspection of vessels at sea. The term is now used more broadly to encompass prevention and deterrence, and it refers to measures implemented by fisheries management organizations to better manage fish stocks through plans of action at the regional and national levels. The United Nations Convention on the Law of the Sea now considers fisheries MCS as the "best hope for preventing, deterring, and eliminating IUU fishing."[3]

According to the UN FAO, there are three elements of MCS[4]: the *air*, *land*, and *sea*. For fisheries management, *air* MCS means the use of satellite technologies to monitor the activities of both national and foreign fishing vessels in the

Box 5.1 Monitoring, Control, and Surveillance

The United Nations Food and Agriculture Organization (UN FAO) defines the components of MCS as follows:

> Monitoring—the continuous requirement for the measurement of fishing effort characteristics and resource yield;
>
> Control—the regulatory conditions under which the exploitation of the resources may be conducted; and
>
> Surveillance—the degree and types of observations required to maintain compliance with the regulatory controls imposed on fishing activities.[1]

Note

1. Flewwelling, P. (2004). *An Introduction to Monitoring, Control and Surveillance Systems for Capture Fisheries.* FAO Fisheries Technical Paper 338. http://www.fao.org/DOCREP/003/V4250E/V4250E00.htm#toc.

country's waters to detect possible violations. Using *land* MCS means conducting port inspections to prevent illegal fish landings and illegal transshipments, as well as controlling access to marine resources by, for example, closely monitoring foreign access agreements (with foreign fishing fleets) or licensing of national fleets. Using *sea* MCS means placing observers onboard fishing vessels or conducting inspections at sea. In this chapter, we will explore two of these MCS components: *sea* and *land.*

To examine "MCS at sea," this chapter looks at the countries' "observer schemes," "vessel monitoring schemes," and "the number of patrol vessels." To study the role of "MCS at land," this chapter examines how these 53 countries performed on "catch inspection schemes" and "control of access to stop illegal fishing" measures. To better understand what these "schemes" mean, we will begin with a brief description of each of these programs.

Observer Schemes

This refers to a program that ensures that fishing vessels operate in compliance with conservation and management measures established by a country. The UN FAO Code of Conduct for Responsible Fisheries[5] encourages countries to use observers (i.e., fisheries inspectors) onboard fishing vessels whose primary functions will include generating data for fishery science and compliance purposes. For fishery science purposes, observers collect data on total catch and effort, which also include bycatch and discards, as well as the biological sampling of catches. For the purposes of compliance with fisheries laws and regulations, observers may be authorized to validate logbooks (i.e., validate information recorded by fishing vessels), examine gear used, and inspect other relevant paperwork. Such information serves an important goal of ensuring effective fisheries management, as well as helps detect unreported fishing. In most countries, observer schemes are legally imposed programs, and fisheries observers are placed onboard vessels by either the fisheries management organization or the coast guard.

Vessel Monitoring Schemes

The UN FAO also encourages countries to implement vessel monitoring schemes in accordance with relevant national, regional, and international standards. These programs require that vessels carry vessel monitoring systems (VMS) (i.e., tracking devises) onboard. Through VMS, the country can closely track both the activities of licensed fishing vessels operating within its EEZ and the overseas activities of the vessels flying its flag to ensure compliance with international regulations. There are two types of VMS. The more conventional type relies on tracking vessel movements through radio, aerial, and surface surveillance, as well as land-based radar and sea-based sonar. This type of VMS is more local in coverage. The more sophisticated type of VMS is satellite based, and it relies on satellite technologies such as Inmarsat, Argos, and Euteltracs to transmit and receive signals from fishing vessels. Countries use these data to not only track the locations of the vessels in larger range but also calculate the speed and the course of these vessels.[6]

Catch Inspection Schemes

These programs refer to the enforcement of fisheries laws and regulations and include several key components: detection, approach, boarding, inspection, and disembarkation.[7] Inspections may involve routine boarding or inspection or boarding with suspected violation. The at-sea inspection involves gathering data in order to determine compliance with the terms of the license and fisheries regulations. Catch inspection schemes may also involve inspections of catch at ports and the verification of catch through logbooks and other documents. The inspection procedures at ports include vessel identification, inspection of authorization to fish (i.e., license) and other documentation, and examination of fishing gear and the fish caught. Vessel identification includes the verification of the validity of the vessel identity documents (through appropriate contacts with flag states and regional fisheries management organizations) that include the vessel's external identification number and international radio call signs.

Control of Access to Stop Illegal Fishing

This measure includes having an effective licensing system in place for fishing vessels and incorporates the following measures: (a) authorization to fish,

(b) establishment of records of fishing vessels, (c) maintenance of logbooks for licensed vessels, (d) authorization and control of transshipment activities, (e) regulating fishing access to the EEZ, and (f) denying licenses to known IUU fishing vessels.

How Are These MCS Measures Expected to Reduce the Risk of Illegal Fishing?

It is assumed that countries with strong MCS measures will experience notably lower degrees of illegal fishing activities within their waters. This is because these measures, if implemented systematically, would increase the likely illegal fishers' perceived risk of being caught. The fisheries MCS measures outlined earlier are, in effect, the "screening exits," "reducing anonymity," "utilizing place managers," and "controlling access to facilities" situational crime prevention techniques. It is also expected that rational offenders would avoid areas with strong formal surveillance.

THE ROLE OF PORTS OF CONVENIENCE AND ATTRACTIVE RESOURCES IN INCREASING THE RISK OF ILLEGAL FISHING

Ports of Convenience Revisited

Researchers at the Pew Environmental Group undertook a study where they examined the movements of illegal fishing vessels during a six-year period.[8] Their findings revealed that these vessels (a total of 58) made 425 visits to 140 ports in 71 countries. The study also provided more in-depth information on 94 visits in 32 countries. These 32 countries were those that had at least four visits during the study period. An examination of these port visits revealed that 10 ports (7 percent of the ports visited) accounted for 155 visits (37 percent of all port visits by illegal fishing vessels). Table 5.1 summarizes the results for these 10 ports (henceforth, "POCs"). In this chapter, we will focus on measuring the countries' proximity to these 10 POCs and estimate these countries' vulnerability to illegal fishing[9] due to this proximity.

Resource Attractiveness

An evaluation of resource "attractiveness" is done by examining the availability of species that are highly commercial internationally within each of the 53 countries' EEZs. Researchers have suggested that countries that have more such resources are more vulnerable to illegal fishing, as they possess more valuable targets that are attractive "hot products."[10] Table 5.2 lists these highly commercial species that are found in 10 or more countries.

Table 5.1 Number of Detected Port Visits by Illegal Fishing Vessels, 2004–2009

Country	Name of Port	Number of Visits	Percentage of Visits	Cumulative Percentage of Visits	Cumulative Percentage of Ports
Singapore	Singapore	32	7.5	7.5	0.7
Ecuador	Bahia de Manta	16	3.8	11.3	1.4
Ukraine	Sevastopol	16	3.8	15.1	2.1
Mauritania	Nouadhibou	15	3.5	18.6	2.9
Russia	Kaliningrad	15	3.5	22.1	3.6
Colombia	Cartagena	14	3.3	25.4	4.3
Germany	Rostock	14	3.3	28.7	5.0
South Korea	Pusan	12	2.8	31.5	5.7
Ghana	Tema	11	2.6	34.1	6.4
Spain	Las Palmas	10	2.4	36.5	7.1
Total		**155**	100.0	100.0	100.0

Table 5.2 List of Highly Commercial Fish Species That Are Found in 10 or More Countries

Common Name	Scientific Name	Number of Countries	Common Name	Scientific Name	Number of Countries
Rainbow trout	*Oncorhynchus mykiss*	27	Haddock	*Melanogrammus aeglefinus*	12
Common carp	*Cyprinus carpio*	25	Largehead hairtail	*Trichiurus lepturus*	12
Skipjack tuna	*Katsuwonus pelamis*	24	Saithe	*Pollachius virens*	12
Chub mackerel	*Scomber japonicus*	19	Atlantic horse mackerel	*Trachurus trachurus*	11
Yellowfin tuna	*Thunnus albacares*	19	Blue whiting	*Micromesistius poutassou*	11
Atlantic mackerel	*Scomber scombrus*	16	Nile tilapia	*Oreochromis niloticus*	11
Atlantic cod	*Gadus morhua*	15	South American pilchard	*Sardinops sagax*	11
Atlantic herring	*Clupea harengus*	15	Barramundi	*Lates calcarifer*	10
Bigeye tuna	*Thunnus obesus*	15	Common sole	*Solea solea*	10
Flathead grey mullet	*Mugil cephalus*	15	European plaice	*Pleuronectes platessa*	10
Albacore tuna	*Thunnus albacares*	14	European sprat	*Sprattus sprattus*	10
Kawakawa	*Euthynnus affinis*	13	Fourfinger threadfin	*Eleutheronema tetradactylum*	10
Atlantic salmon	*Salmo salar*	12	Narrow-barred Spanish mackerel	*Scomberomorus commerson*	10
Common dolphinfish	*Coryphaena hippurus*	12	Whiting	*Merlangius merlangus*	10

Box 5.2 Expectations Deriving from the Theoretical Framework of Environmental Criminology

Countries with strong MCS measures at sea (i.e., effective *observer schemes, vessel monitoring schemes*, and *more patrol vessels*) are less likely to experience high levels of illegal fishing within their EEZs.
Countries with strong MCS at land (i.e., effective *control of access* and *catch inspection schemes*) are less likely to experience high levels of illegal fishing within their EEZs.
Countries within close proximity to top 10 ports of convenience will experience significantly more illegal fishing within their EEZs.
The number of (internationally) highly commercial fish species within a country's EEZ will be an important factor in fishers' decision to engage in illegal fishing.

How Are These Factors Expected to Increase the Risk of Illegal Fishing?

Research on illegal fishing, as discussed in Chapter 3, has identified the importance of POCs as facilitators of illegal fishing. Regardless of the local measures taken by a country, it may still be vulnerable to illegal fishing because of its proximity to a port with lax regulations in place. These types of ports allow the illegal fishing vessels to conceal their catches by unloading and transferring these via other methods to target destinations. Therefore, on the basis of the framework of the rational choice perspective, fishing within a country's EEZ and unloading the fish in a POC would reduce the risk of being detected at landing. Consequently, access and availability of such ports within a convenient distance from the fishing grounds of a country would make that country a vulnerable target for illegal fishing.

Additionally, one would expect illegal fishing to be highest in areas where there are highly sought-after resources. These resources can be "attractive" for those fishing illegally because they are in high demand in international markets and, consequently, will be easily disposed of (see Box 5.2 for a list of theoretical assumptions that will be tested later in this chapter).

WHAT DO THE NUMBERS SUGGEST?

Before numbers are examined, a brief mention of the data sources is necessary.

Data Sources

Source 1: University of British Columbia

Eleven years past the 1995 UN Code of Conduct for Responsible Fisheries,[11] scholars from the University of British Columbia conducted a study where they evaluated countries' compliance with the provisions outlined in this code and their overall performance on their efforts to deal with different fisheries-related issues, such as "management of stocks, fleets, and gear"; "socio-economics"; and "monitoring, control, and surveillance."[12] Within the "monitoring, control, and surveillance" evaluation field, they examined such important factors as the effectiveness of "observer schemes," "vessel monitoring schemes," "catch inspection schemes," and "control of access to stop illegal fishing" and the "degree of illegal fishing" and ranked these countries' performance on these measures on a scale of 0–10.[13]

Source 2: The Sea Around Us Project

This research project, launched in 1999, was also the work of scholars at the University of British Columbia, who built a publicly available database (see http://searoundus.org) that provides information on "increasingly sophisticated analyses of trends in global fisheries that allow policymakers and fisheries managers to make more accurate and informed decisions."[14] The EEZ-specific data that can be accessed through this database allow for in-depth analyses of how attractive these 53 countries' EEZs are in terms of the commercially significant resources they offer, that is, fish species that are highly commercial in international markets.

Source 3: U.S. Naval Institute

The Naval Institute *Guide to* Combat Fleets of the World[15] uses an extensive list of unclassified sources, such as *The Almanac of Seapower*, Defense News, *Flight International*, *Intelligence, Surveillance & Reconnaissance Journal*, and *Marine News*, to compile an inventory of the countries' coastal and air defense resources. Most relevant to fisheries management is the number of patrol vessels that each country has, and this information was extracted from this guide.

Which Countries Have a High Degree of Illegal Fishing?

Table 5.3 shows the degree of illegal fishing occurring in the waters of these 53 countries. A striking number of countries have illegal fishing within their EEZs. In about 42 percent of these countries, the degree of illegal fishing is "very high." Additionally, 16 countries have a "high" degree of illegal fishing. Cumulatively, a total of 38 countries, comprising about 72 percent of the 53 important fishing countries, have to deal with serious illegal fishing activities occurring in their EEZs.

To further quantify the vulnerability of the countries that have a "very high" degree of illegal fishing, Figure 5.1 demonstrates these countries' performance scores in more detail, where a score of "10" means that there are "as many as or more than legal vessels" fishing illegally in the country's waters, and a score

Table 5.3 The Degree of Illegal Fishing within the EEZs of 53 Countries

Degree of Illegal Fishing	Number of Countries	Percentage of Countries
Very high	22	41.5
High	16	30.2
Medium	13	24.5
Low	2	3.8

Figure 5.1 Countries Where Illegal Fishing Activity Is "Very High" and Their Vulnerability Scores

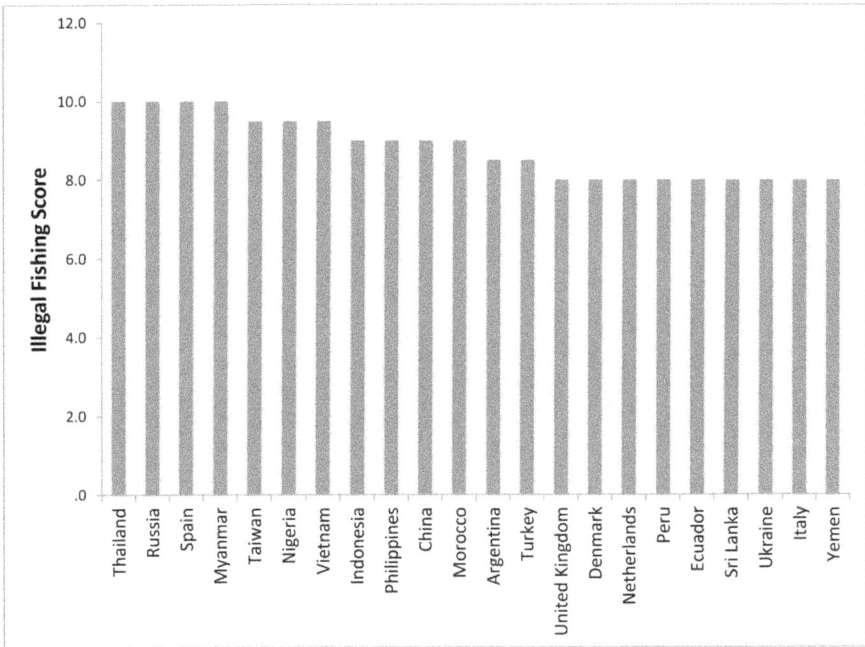

of "8" indicates that there is a "great deal of or half as much as legal vessels" fishing illegally in the country's waters. Four countries, including Thailand, Myanmar, Russia, and Spain, scored the highest among the 22 top illegal fishing countries, and more than half of these countries had a score of "9" or above.

What Is the Geographic Dimension of the Prevalence of Illegal Fishing?

While most countries are vulnerable to illegal fishing to different degrees, this may be explained by the countries' geography. For example, some countries

Figure 5.2 Geographic Distribution of the Patterns of Illegal Fishing within 53 EEZs

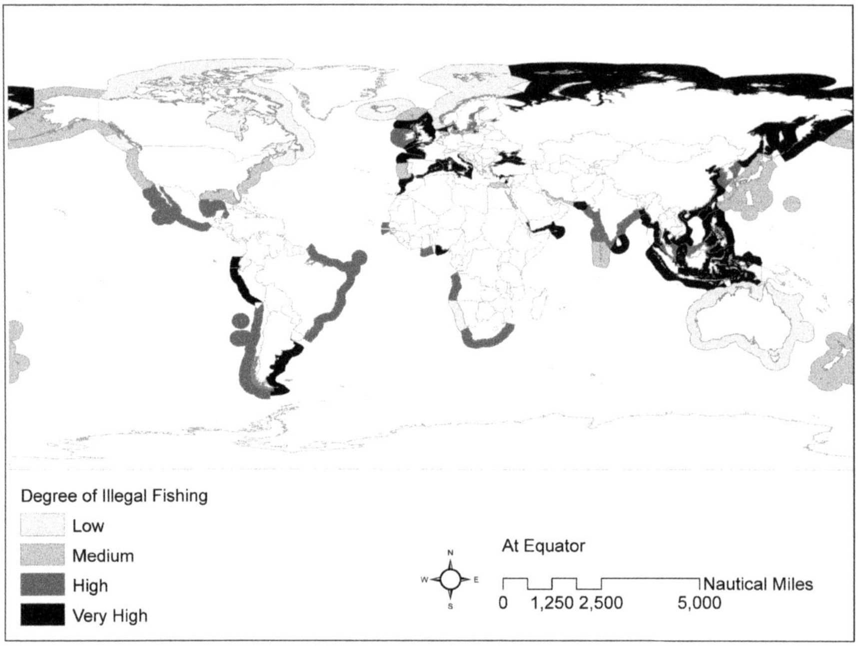

where illegal fishing is high are likely to be within a close geographic proximity to other countries that have similar characteristics. Turning to Figure 5.2, one can see some distinct geographic patterns.

Illegal fishing is "high" in all continents, except for Australia, where its degree is "medium." However, some continents have notably more illegal fishing than others, including South America, Africa, Europe, and Asia. In South America, illegal fishing is "very high" in the northwestern and southeastern coasts, within the EEZs of Ecuador, Peru, and Argentina. In Africa, illegal fishing is particularly concentrated along the Western African coastline, with the exception of Namibia, where the degree of illegal fishing is "low." One can see similar patterns in Europe, where nine countries experience a "very high" degree of illegal fishing, including Spain and Italy in the south, Denmark and Ireland in the northeast, the United Kingdom in the north, France and the Netherlands in the west, and Russia and Ukraine in the east. Unlike South America, Africa, and Europe, where countries experience different degrees of illegal fishing, countries in Southeast Asia are equally vulnerable to "very high" degrees of illegal fishing. These countries—namely, Indonesia, Vietnam, the Philippines, Myanmar, Malaysia, Thailand, Taiwan, and Bangladesh—form a geographic "hub" for a "very high" degree of illegal fishing.

Is the Degree of Illegal Fishing Correlated with MCS Measures in Place?

MCS at Sea: Observer Schemes

Figure 5.3 shows the degree of illegal fishing and the effectiveness of the countries' observer schemes. The countries that performed poorly on their observer schemes are the ones that have a high degree of illegal fishing. In fact, 13 countries scored "0" on the "observer schemes" measure, indicating that these countries had no such programs in place.[16]

MCS at Sea: VMS

Figure 5.4 shows how the countries' vessel monitoring schemes are related to the degree of illegal fishing activities occurring within their waters. While, overall, countries had better vessel monitoring schemes than observer schemes in places, the strength or absence of such schemes was again correlated with the degree of illegal fishing within their waters. A total of eight countries with no such programs or those with weak programs in place had much more illegal fishing than the countries with stronger vessel monitoring schemes in place.[17]

MCS at Sea: Patrol Surveillance

Figure 5.5 shows how the number of countries' "patrol vessels per 10,000 sq km of EEZ" is related to the degree of illegal fishing within their EEZs. While generally it indicates that having fewer patrol vessels means experiencing higher degree of illegal fishing, what is most striking is that 19 countries had less than one patrol vessel per 10,000 sq km of EEZ, and an additional 11 countries had only one. Cumulatively, 57 percent of the countries (30 countries) had one or fewer patrol vessels per 10,000 sq km of EEZ. With the exception of Thailand, which had 10 vessels, the remaining countries had 8 or less such vessels patrolling their EEZs.

MCS at Land: Catch Inspection Schemes

Figure 5.6 displays the relationship between the degree of illegal fishing and the countries' catch inspection schemes. Similar to the patterns observed earlier, weak or nonexistent catch inspection programs were associated with the higher degree of illegal fishing within the countries' EEZs. Only Nigeria and North Korea scored "0" on this measure, indicating no such programs in place. Other countries had some kind of catch inspection program, albeit the performance of most countries on this measure was average.

Figure 5.3 The Relationship between the Degree of Illegal Fishing and the Effectiveness of Countries' Observer Programs

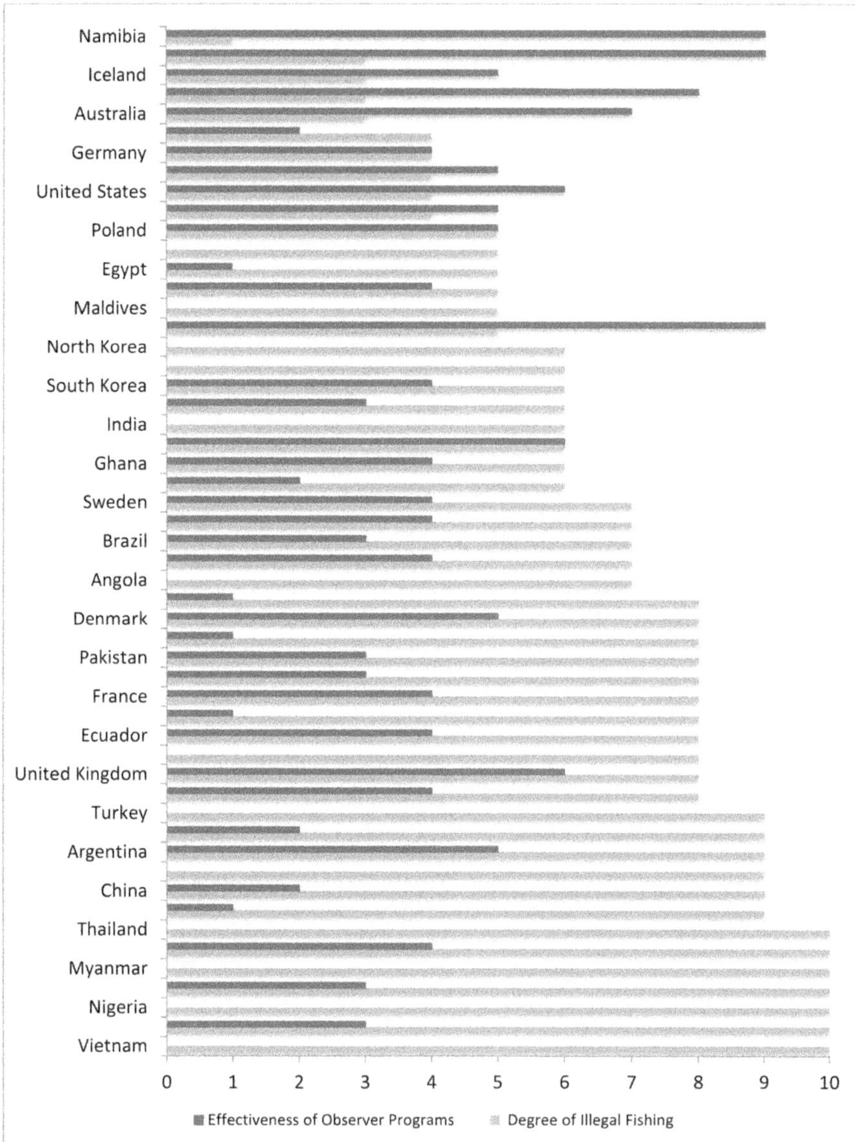

Figure 5.4 The Relationship between the Degree of Illegal Fishing and the Effectiveness of Countries' Vessel Monitoring Programs

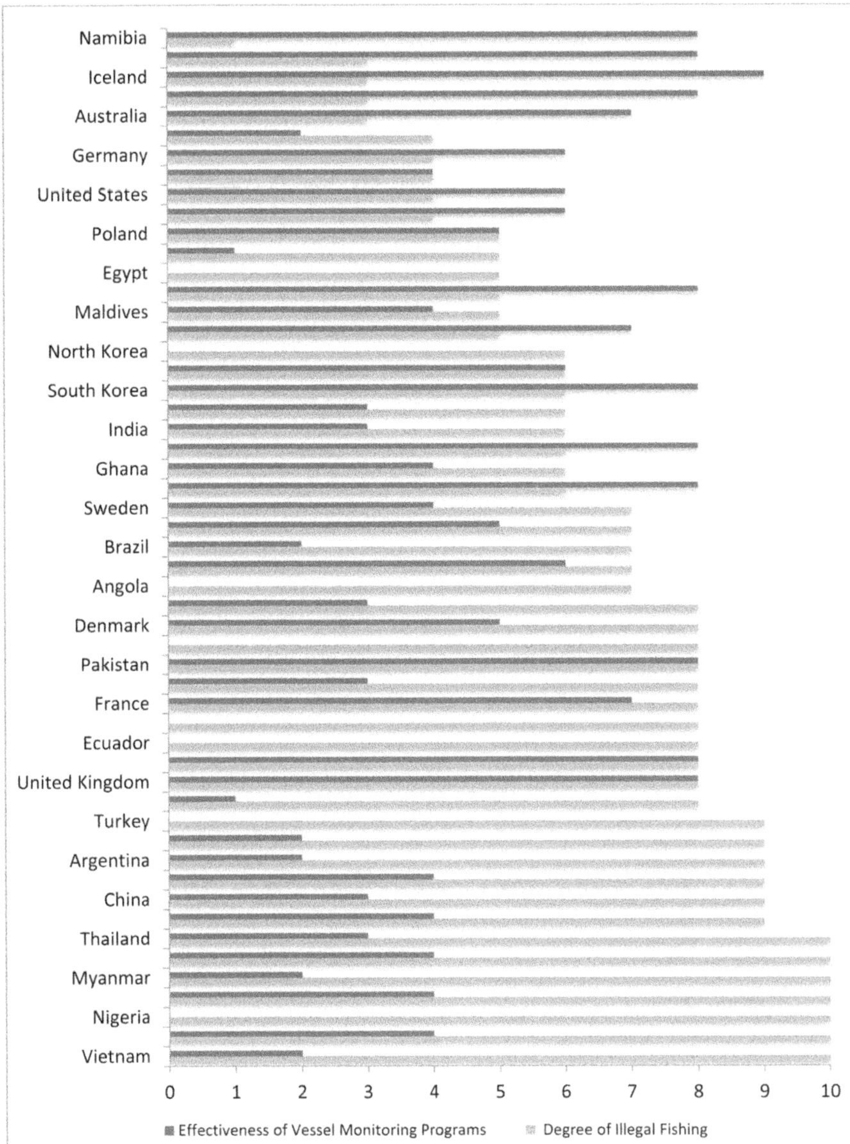

Figure 5.5 The Relationship between the Degree of Illegal Fishing and the Number of Patrol Vessels per 10,000 sq km of EEZ

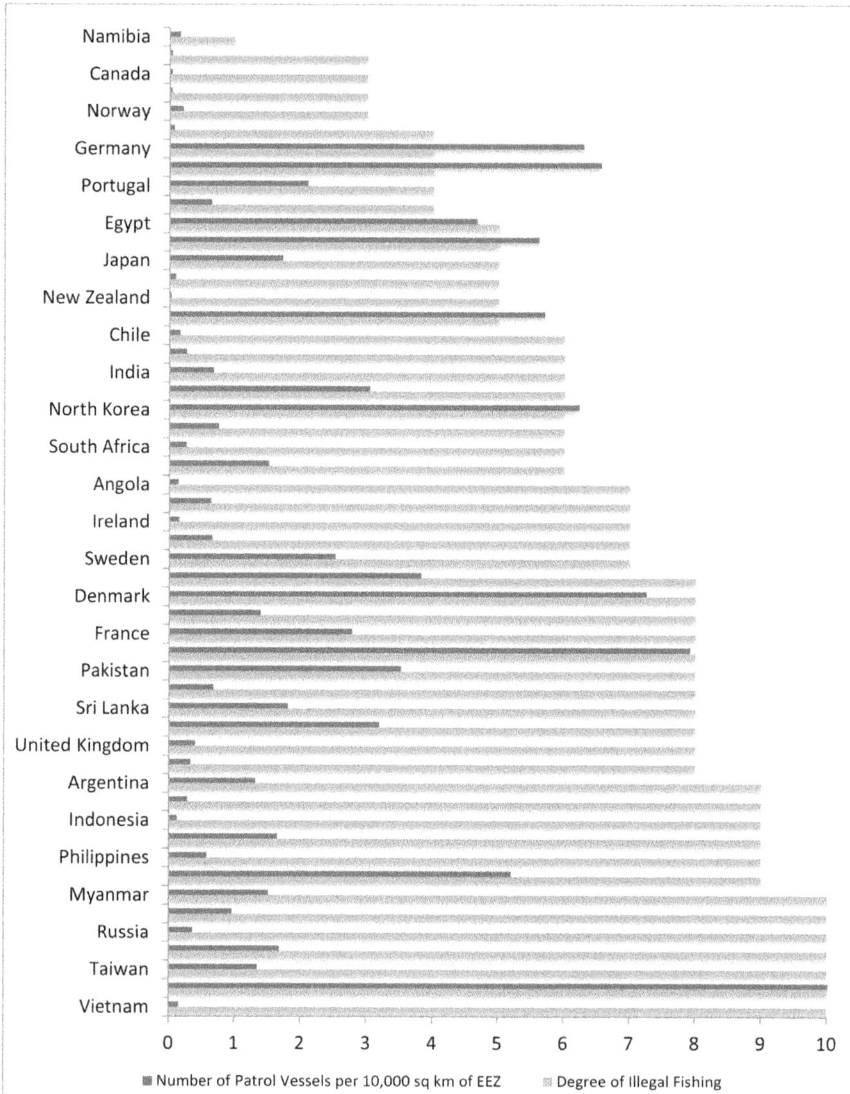

Figure 5.6 The Relationship between the Degree of Illegal Fishing and the Effectiveness of Catch Inspection Programs

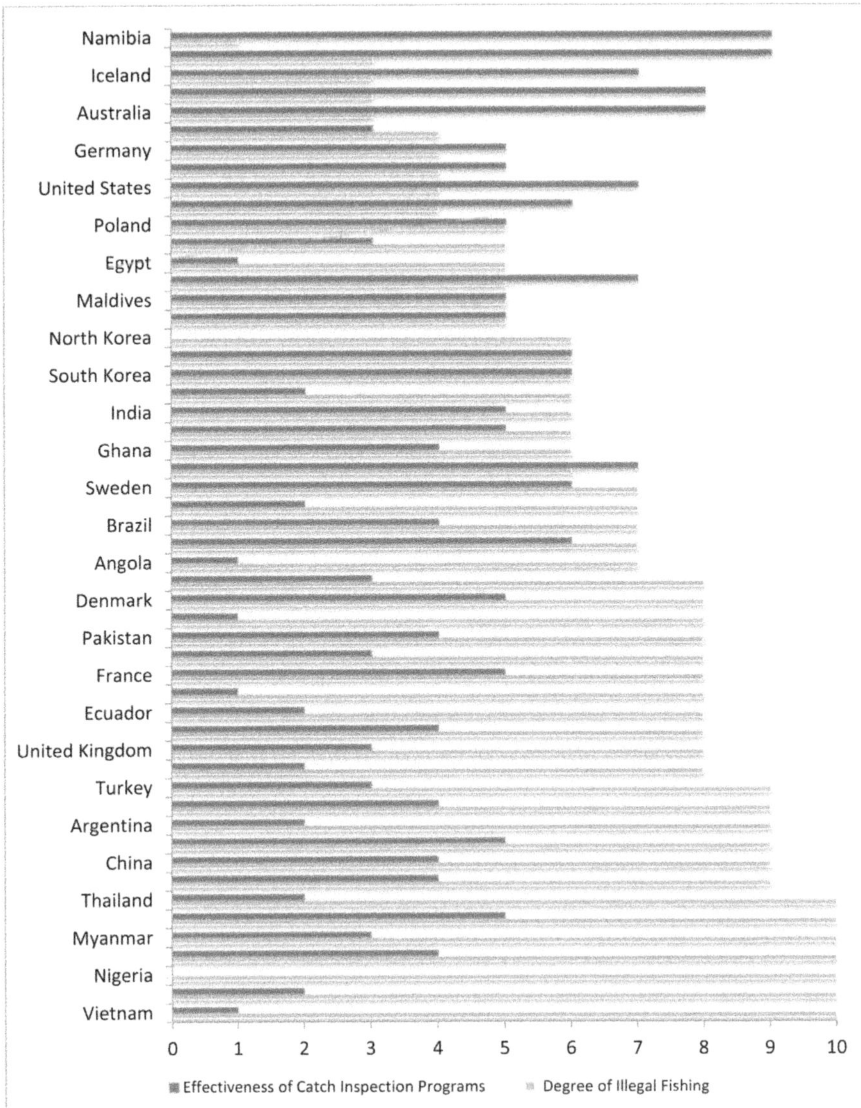

Figure 5.7 Effectiveness of Access Control Measures to Stop Illegal Fishing and the Degree of Illegal Fishing in Countries' EEZs

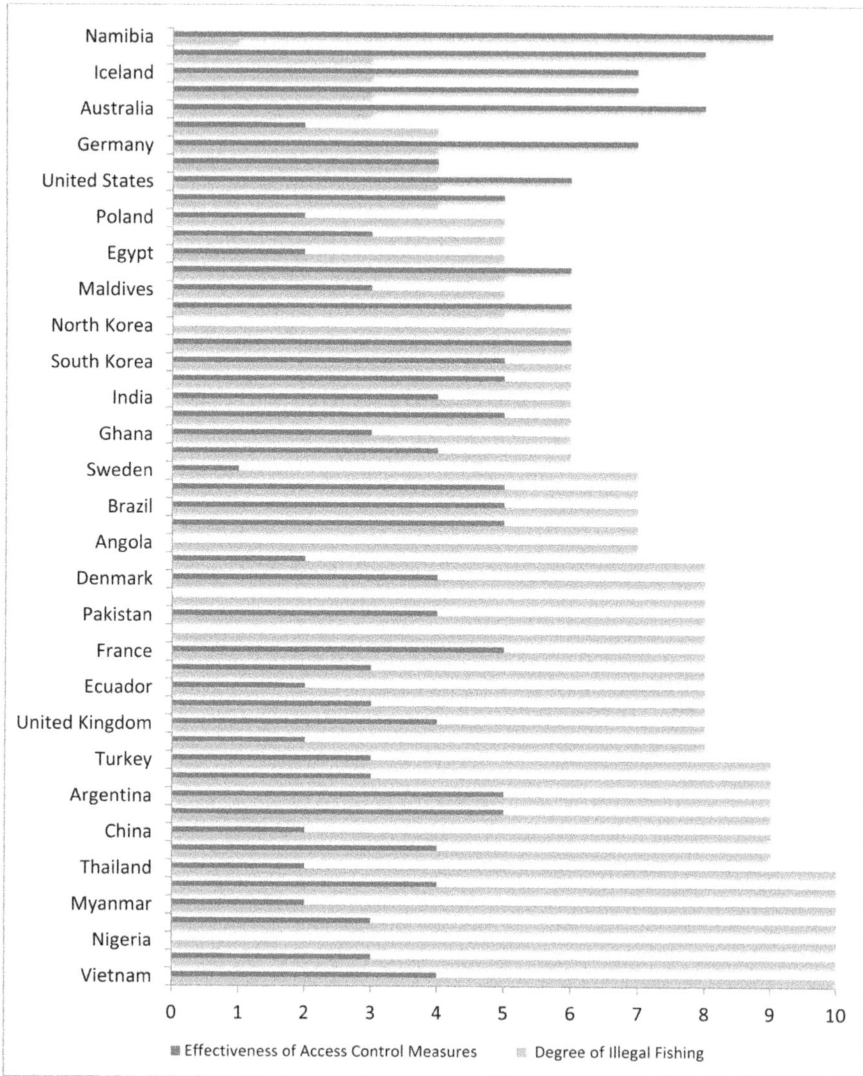

MCS at Land: Control of Access to Stop Illegal Fishing

Figure 5.7 shows how the efforts to "control access to stop illegal fishing" are related to the degree of illegal fishing in the countries' waters. Poorly performing countries had a higher degree of illegal fishing due to the absence of poor "access control measures in place to stop illegal fishing." Five countries—namely,

Ukraine, Bangladesh, Angola, North Korea, and Nigeria—earned a score of "0," indicating complete absence of such measures.

Overall, much fewer countries have performed almost perfectly on their MCS efforts, and the "best" score was "9" on a 10-point scale. Two countries—in particular, Nigeria and North Korea—lack all MCS measure, receiving a score of "0" on all four MCS. Other countries (e.g., Angola) lacked a majority of these measures, receiving a score of "0" on three of these measures, and Bangladesh lacked half of these measures, receiving a score of "0" on two of these measures. Importantly, the majority of the countries that performed poorly on these measures are developing countries. Overall, on all these measures, the 53 countries' performance on any of these MCS measures is fairly poor. Same countries experience a high degree of illegal fishing within their waters.

Is the Degree of Illegal Fishing within a Country's EEZ Related to the Country's Relative Proximity to the Known POCs or the Number of Attractive Resources?

Access to Ports of Convenience

Table 5.4 shows how many of the top 10 riskiest POCs are accessible to each of these 53 countries. Only about one-third of the countries do not have access to any of these ports available within a close geographic proximity[18]; however, the remaining 70 percent of the countries are close to at least one such port.

An examination of the accessibility of these POCs in Figure 5.8, where access to at least one such port is geographically examined, reveals two important patterns. First, countries in South and Southeastern Asia, as well as most countries in Europe, overwhelmingly have access to at least one such port. Second, and importantly, accessibility to these ports mirrors the locations where illegal fishing is fairly high.

Table 5.4 Number of Ports of Convenience Available within 1,500 nm from Each of the 53 Countries

Number of Available Ports of Convenience	Number of Countries within 1,500 nm to Ports of Convenience	Percentage of the Total Available Ports of Convenience	Cumulative Percentage
2	15	28.3	28.3
1	22	41.5	69.8
0	16	30.2	100
TOTAL	**53**	**100**	

Figure 5.8 Accessibility of Top 10 Riskiest Ports of Convenience to the 53 Fishing Countries

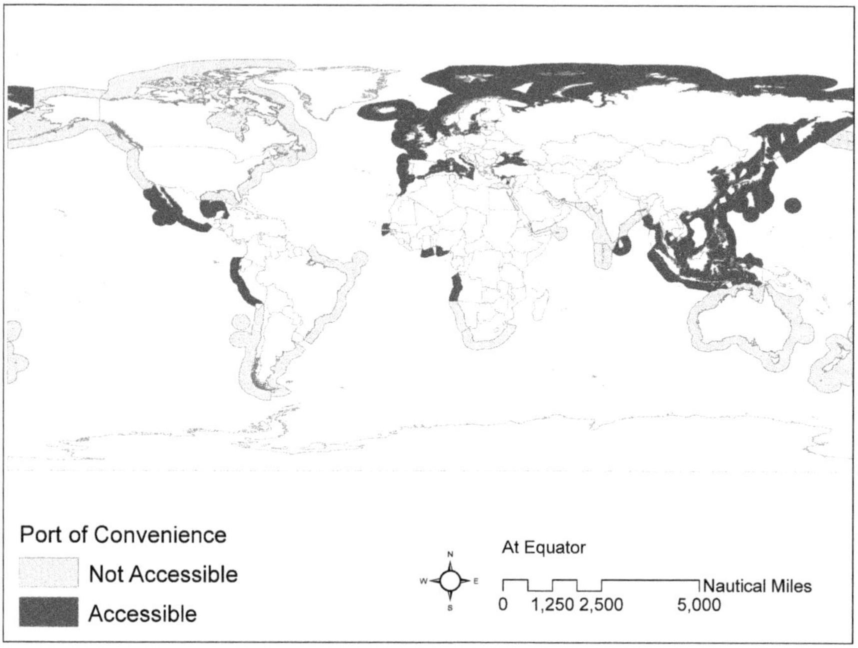

Singapore is a particularly interesting port to examine, as 32 out of the total of 155 illegal port visits (i.e., 21 percent) to these 10 riskiest POCs were made to this port alone (during 2004–2009). A further analysis of the countries that are in close proximity to Singapore—namely, Thailand, Myanmar, Vietnam, Indonesia, the Philippines, and Malaysia—revealed that except for Malaysia, the remaining countries scored "very high" on the degree of illegal fishing taking place within their territorial waters.

Attractive Resources

When examining the number of species that are highly commercial internationally, it was discovered that these countries had an average of 23 highly commercial species, and some countries had as many as 55. The countries that had 23 or more such species are shown in the graph in Figure 5.9.

The graph also shows these countries' score on the degree of illegal fishing within their territorial waters. A total of 13 (i.e., 25 percent) countries had 23 or more species that were highly commercial internationally, and a majority of

Figure 5.9 Countries with 23 or More Highly Commercial Species

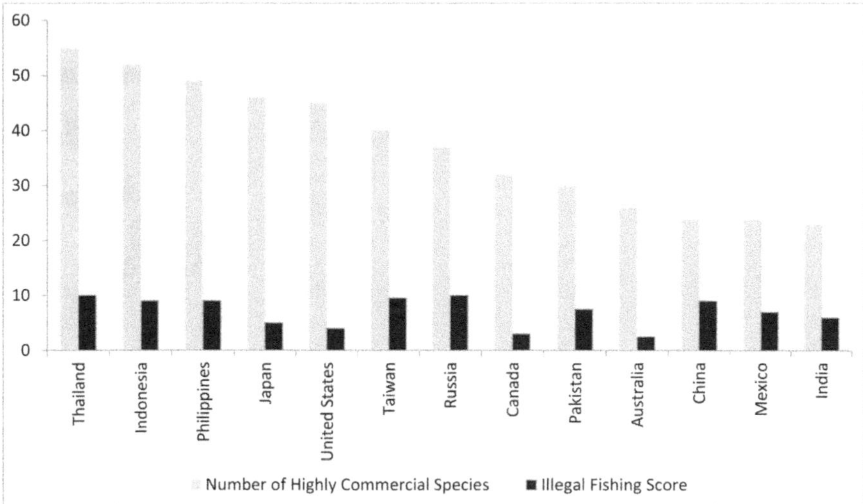

these countries (62 percent) scored "very high" on the degree of illegal fishing prevalent in their territorial waters. Importantly, the majority of these countries are in Southeast and East Asia. Thailand, the country that has the most number of commercially significant species ($N = 55$), was among the four countries with a score of "very high" on the degree of illegal fishing. Overall, the presence of these "attractive" fish species is one of the most significant drivers of illegal fishing activity within a country's EEZ.

Are the Patterns of Distribution of Attractive Resources Geographically Correlated with Those of Illegal Fishing?

To examine whether the species that are highly commercial internationally are geographically clustered around areas with "very high" degrees of illegal fishing, we turn to Figure 5.10.

Recalling the geographic "hub" of illegal fishing, it is interesting to see what Figure 5.10 reveals: countries that have the highest number of such resources form the same "hub" in Southeast Asia. Similarly, European countries that have relatively high number of "attractive" resources (between 16 and 30) form a cluster where illegal fishing is high. Surprisingly, South American countries do not have as many attractive resources as other countries despite the fact that these countries are among those that experience "very high" degrees of illegal fishing.

Figure 5.10 Number of Highly Commercial Species

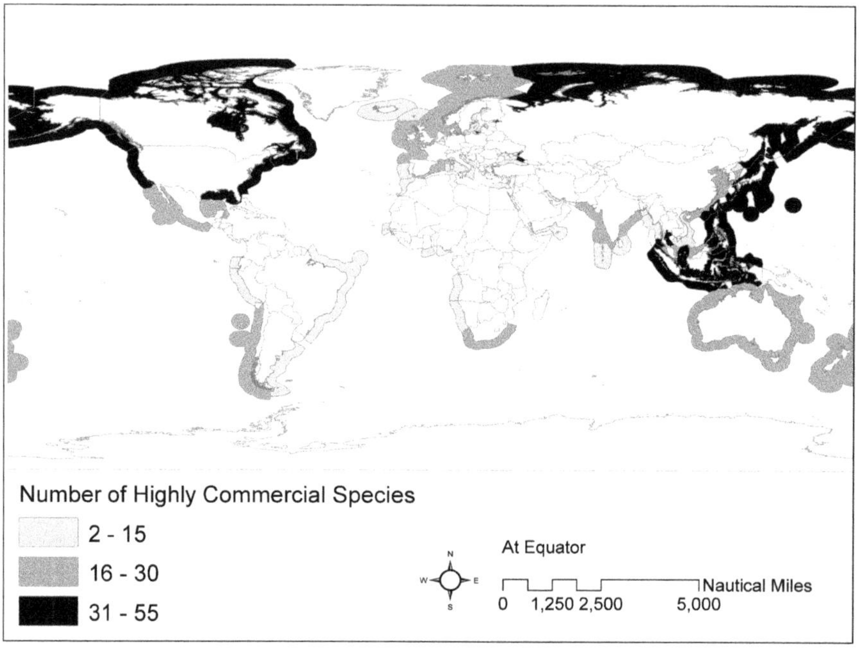

CONCLUDING REMARKS

Data examined in this chapter revealed much of what was expected about the patterns of illegal fishing. About 70 percent of the world's most important fishing countries have serious illegal fishing problems—by itself an alarming finding. The same 70 percent of countries are in close proximity to the top 10 riskiest POCs. The most problematic of these ports, Port Singapore, is geographically situated around some of the most vulnerable countries where illegal fishing is rampant, not to mention its overall convenient location in the Southeast Asian region that is a known hot spot area for illegal fishing. None of the countries' patrol vessel capacity is sufficient enough to reduce the risk of illegal fishing, considering the greatest majority of these countries have eight or less such vessels per 10,000 sq km of EEZ area. A great majority of these countries have a large number of the world's most sought-after fish species within their waters, and these species' distributions also cluster in the parts of the world where illegal fishing is a problem. In fact, researchers have built more complex models using the data presented in this chapter to examine the relationship between these factors and the degree of illegal fishing, and they have come to the exact same conclusions about what increases or reduces the risk of illegal fishing.[19] Therefore, fisheries MCS at sea and at land, indeed, remain the "best hope for preventing, deterring, and eliminating IUU fishing."[20]

Chapter 6

What Has Been Done to Address the Problem of Illegal Fishing?

MONITORING AND SURVEILLANCE

The United Nations' Regulatory Blueprints (Technique # 21: Set Rules)

Legally Binding Fisheries Laws

"A Constitution for the Seas": The 1982 United Nations Convention on the Law of the Sea (UNCLOS) The first UN Conference on the Law of the Sea took place in 1958, during which four conventions were adopted: the Convention on the Territorial Sea and Contiguous Zone, the Convention on the Continental Shelf, the Convention on the High Seas, and the Convention on Fishing and Conservation of Living Resources of the High Seas. After the adoption of these conventions, however, some important issues were left unresolved, such as those pertaining to the rights to the resources of the deep seabed and the breadth of the territorial seas.[1] As a result, a series of UN conferences on the law of the sea followed between 1974 and 1982. The UNCLOS was eventually opened for signature in 1982 in Montego Bay, Jamaica, and signed by 155 states. In 1994, the UN Secretary General Boutros Boutros-Ghali initiated consultations that resulted in the Agreement Relating to the Implementation of the Part XI of the UNCLOS of 1982, which then went into force in 1996 and received 138 ratifications. To date, 168 states, which include 167 states and the European Union, have ratified the UNCLOS.

The UNCLOS provides rules on how to regulate oceans and seas under the important principle of the *equitable and efficient utilization* and the *conservation* of living resources. These rules provide a framework to nation-states to develop conservation and management regulations relating to the marine resources within their exclusive economic zones (EEZs), such as those pertaining to the licensing of fisheries; fishing vessel, equipment, and gear types; and the quotas on and types of the species to be caught; as well as regulations that govern foreign fishing vessels' access into their EEZs. Similarly, the UNCLOS outlined measures that regional fisheries management organizations (RFMOs) can adopt to deal with the management of the marine resources in the high seas[2]; as well as manage transboundary and straddling[3] fish stocks and highly migratory species. In essence, the UNCLOS is a blueprint for states to develop effective ocean conservation and prevention measures.

Agreement to Promote Compliance with International Conservation and Management Measures by Fishing Vessels on the High Seas (FAO Compliance Agreement) In an effort to address the problem of *reflagging*, a practice that is still widely used by illegal fishing vessels to avoid compliance and responsibility, as well as to deal with other compliance issues pertaining to the vessels fishing beyond the nation-states' EEZs, the UN Food and Agriculture Organization (FAO) Compliance Agreement was adopted in 1993, and it went into force in 2003. The agreement outlines specific measures that nation-states must adopt in order to deal with their fishing vessels' compliance with international conservation and management regulations. For example, states must ensure that none of their vessels are fishing on the high seas without authorization. These states must also adopt guidelines to prevent their vessels from reflagging under the flags of convenience. Lastly, these states must maintain a record of all national fishing vessels fishing in the high seas, as well as cooperate with other states for the exchange of information related to the fishing operations of their national vessels.

UN Fish Stocks Agreement The 1995 UN Fish Stocks Agreement, which entered into force in 2001, primarily deals with the conservation and management of *straddling and highly migratory fish stocks in the high seas*. The agreement outlines requirements that concern the compliance, catch verification, and reporting practices. For example, the nation-states are required to mark their fishing vessels and gear, impose vessel monitoring requirements, and implement observer programs, as well as adopt and enforce measures relating to the boarding and inspection of vessels at their ports. A significant aspect of the agreement is its requirement that states establish regional or subregional RFMOs as an effort to conserve and manage these shared fish stocks.

Agreement on Port State Measures to Prevent, Deter, and Eliminate Illegal, Unreported, and Unregulated Fishing (Port State Measures Agreement) The 2009 FAO Port State Measures Agreement (PSMA) contains provisions that integrate *measures to be implemented at states' ports*. Prior to this agreement, states were free to develop and implement regulations that related to the inspection of vessels at their ports, the information required from vessels entering into these ports, and the types of penalties imposed for violations. As such, states had differential standards in terms of these measures. The agreement, while reiterating the sovereignty of port states in adopting measures designed to combat illegal fishing, nevertheless contains provisions that harmonize port state regulatory practices.[4]

To *ensure compliance at ports*, PSMA requires that states deny landings and transshipment to fishing, refrigerated cargo, and carrier vessels, as well as support ships, if the states determine that these vessels have been listed as illegal, unreported, and unregulated (IUU) fishing vessels by RFMOs or they carry the flag of a known flag-of-convenience state. If allowed landing at ports, PSMA requires that the port states implement robust inspection procedures at port on arriving fishing vessels. Prior to arriving at a port, the vessels should be required to provide notification for port access, which would include information such as (a) the vessel's identification number, (b) fishing license, (c) fishing trip and species caught, and (d) the vessel's monitoring system. Additionally, if port inspectors determine that the vessel has been engaged in illegal fishing, they should be required to communicate with the flag state to relay that information.

Voluntary Agreements

FAO Code of Conduct for Responsible Fisheries (FAO Code of Conduct) The FAO Code of Conduct, which was adopted on October 31, 1995, proposes recommendations that the states can use to *minimize waste, discards, and negative impacts of fishing*. The code includes provisions that relate to the conservation, management, and development of all fisheries and general principles pertaining to fisheries management, fishing operations, aquaculture development, fisheries integration into coastal area management, and so on. Importantly, the code addresses a range of issues that relate to IUU fishing activities, such as using destructive fishing gear and fishing without authorization.

International Plan of Action on Illegal, Unreported, and Unregulated Fishing (IPOA-IUU) The 2001 IPOA-IUU was designed to help states in their effort to *prevent, deter, and eliminate illegal fishing* either by acting directly or through the relevant RFMO. The IPOA-IUU proposes that port states collect specific information on fishing activities and possibly deny illegal landings and transshipments. In addition, states can impose trade-related measures, such as import bans, and can adopt legislative measures to make it an offense to sell the fish caught by these vessels. The IPOA-IUU also urges coastal states to design effective control and surveillance measures to be implemented in their EEZs.

Interpol and International Law Enforcement Cooperation (SCP Technique # 10: Strengthen Formal Surveillance)

Interpol's Environmental Security Program's Global Fisheries Enforcement team helps international police coordinate their efforts and enhance their

Box 6.1 Sea Shepherd

The Sea Shepherd Conservation Society is an international marine wildlife conservation organization that uses direct action tactics to investigate, document, and stop illegal fishing activities on the high seas.[1] The organization is credited for having a direct impact on the capture of many high-profile vessels. For example, Sea Shepherd's relentless efforts in investigating and gathering intelligence led to the capture and arrest in 2018 by the Peruvian Environmental Prosecutor's Office of the *Damanzaihao*: the world's largest fish factory vessel. Sea Shepherd's efforts were also instrumental in catching the world's most notorious trawler, *Thunder*, in 2015.

Note

1. Sea Shepherd. (2018). *We Are Sea Shepherd.* https://seashepherd.org.

investigative capabilities to handle transnational illegal fishing activities more effectively. With the launch of Project Scale in 2013, which deals with "identifying, deterring and disrupting transnational fisheries crimes,"[5] this international enforcement body seeks to develop different strategies to deal with fisheries crimes worldwide. The team is composed of criminal intelligence officers who have a background in maritime policing, naval services, and fisheries enforcement and other specialists and analysts who have related skills and expertise. These officers work to facilitate the global exchange of law enforcement information among its member states to assist in their investigative efforts. One such tool is Interpol's notices, which are international requests for cooperation or alerts that share critical information from member state police on the movements and activities of individuals and vessels. In relation to fisheries crimes or pertaining to fishing vessels, Interpol has issued Purple and Blue notices. Blue notices generally deal with collecting information about an individual's identity, location, or crime-related activities, while Purple notices provide information on devices, *modus operandi*, and other methods used by criminals. For example, in 2017, Interpol released a Purple notice that details the *modus operandi* of human trafficking in the fisheries sector.

Technologies as Tools to Tackle Illegal Fishing

Satellite Technology (SCP Technique # 8: Reduce Anonymity)

The world's oceans cover an enormous area, and monitoring the movements of illegal fishing vessel activities in the open seas is one of the most daunting tasks. Monitoring, however, can be one of the most promising efforts of successful enforcement if its capabilities can be expanded through different technological advances.

In order to contribute to marine law enforcement efforts, the National Aeronautics and Space Administration, National Oceanic and Atmospheric Administration (NOAA), and the U.S. Intelligence Community have put their remote detection capabilities to use. The *SeaVision* collaboration uses unclassified terrestrial and satellite automatic identification systems (AIS), radar, and satellite imagery to map and track suspected vessels. The U.S. Coast Guard and the NOAA use vessel monitoring systems (VMS), which are satellite-based to track fishing vessel movements in U.S. waters.[6]

In 2015, Secretary of State John Kerry announced the Safe Ocean Network initiative at the Our Ocean Conference in Chile. This is a project led by the U.S. State Department that works with both public and private sector organizations, including various international organizations, industry representatives, different stakeholders, and nongovernmental organizations, to develop a global detection, enforcement, and prosecution capacity to combat illegal fishing. The initiative incorporates emerging technologies and Internet-based tools to enhance collaboration and information sharing.[7]

Oceana, in partnership with Google, SkyTruth, Leonardo DiCaprio Foundation, and the National Geographic Society, among others, has implemented the Global Fishing Watch platform (www.globalfishingwatch.org). This platform collects AIS data from around the world and provides near-real-time maps on fishing vessel activities. This is done by combining Google's software programming, cloud computing, and data management expertise with SkyTruth's hardware that are available to access satellite imagery and other data.

Microsoft's cofounder, the late Paul Allen's Vulcan has launched the SkyLight platform in select parts of the world that uses satellite imagery and data analysis software to aid the law enforcement of developing countries, such as the Pacific Island of Palau and Gabon, where the platform is currently operating, to patrol their EEZs more efficiently.[8]

Blockchain Technology (SCP Technique # 14: Disrupt Markets)

Blockchain technology is one of the most rapidly evolving technologies today, and its use has recently expanded beyond Bitcoin. A pilot project that was launched in 2018 in the Pacific Islands by the World Wildlife Fund in Australia, Fiji, and New Zealand, in partnership with two U.S.-based companies, ConsenSys and TraSeable, and a fishing and processing company based in Fiji, Sea Quest Fiji, Ltd., uses this technology to enhance tuna traceability in the supply chain and increase the transparency of the "journey" of the fish from "bait to plate."[9] This allows close monitoring of the tuna through the use of radio frequency identification (RFID) tags, quick response (QR) code tags, and scanning devices used to collect and disperse information about the tuna's movement at various points along the supply chain. From the moment a tuna is caught, an RFID tag is attached to it, and the information on the tag is uploaded to the blockchain, which triggers the tracking. Once the tuna is landed and processed, the RFID tag is replaced by the QR code tag, which is then linked to the RFID tag via the blockchain. The QR code tag makes it possible to trace the movement of the fish from the moment it is landed and processed to the point of sale. Additional devices, such as near-field communicators (NFCs), are being considered for tracking the fish to the consumer in the future.

The Fishcoin Ecosystem, which was introduced in 2018, is a blockchain-based traceability platform that incentivizes data capturing and transmission among fishers, fish farmers, and buyers. When a transaction occurs between two independent stakeholders, two parallel events happen: (1) the actual real-world asset, for example, the fish being sold, is being traded for fiat currency and (2) the record of that fish (e.g., where it was caught, methods) is unlocked through a "smart contract" by the next receiver in the supply chain, who unlocks this information in exchange for *Fishcoin* tokens. Due to the fact that Fishcoins accrue costs, it is in the best interest of the receiver to ensure the accuracy and representativeness of the seafood asset because if these assets are not verifiable, this may prevent the receiver from selling them along the next supply chain. Each time this data transfer happens among the actors, the data are stored on the blockchain. Fishcoin tokens are designed in the direction *from buyers to sellers*, and, generally, these tokens accumulate near the point of harvest. As a result, seafood producers who make the effort to capture and communicate the data necessary to validate the traceability, legality, safety, and quality of the seafood are those who benefit the most. Buyers (such as restaurants and hotels) benefit from this process of verifying the origin of the harvest in the form of price premiums, while producers can exchange their Fishcoins for airtime top-ups on their prepaid cellphone plans.[10]

Drones (SCP Technique # 7: Assist Natural Surveillance)

In September 2018, the Seychelles Fishing Authority launched the FishGuard pilot project that integrates drones into the daily operations of fisheries patrol of the Seychelles Air Force and the Seychelles Coast Guard. Equipped with artificial intelligence, drones are deployed to short- and long-range surveillance missions to gather information about possible illegal fishing activity within the archipelago's EEZ. The drones are programmed to evaluate the activity independently based on the information collected at sight, and if it is 95 percent confident that the observed activity is illegal, it will notify the relevant local authorities about it and provide detailed information on the time of occurrence, the actual location (with GPS coordinates) where it took place, and other data necessary for the local authorities to decide on the successive course of action.[11]

Box 6.2 Aquatic Drones?

Drone technology developers are considering the idea of developing aquatic drones that can be used to "rule the waves"[1] and catch illegal fishers in action. Liquid Robotics, a subsidiary of Boeing, has developed *Wave Gliders* that use underwater "wings" to collect energy from wave motion or use a propeller powered by solar panels to travel in the water. These gliders are already in use by the Australian and American navies, and Japan's coast guard, who patrol their EEZs and collect a variety of data that include information on locations and movements of submarines, fishing vessels, pollution spots (e.g., from oil spills), water temperature and currents, and so on. The aquatic drones developed specifically for illegal fishing would, for example, take pictures and process them through their onboard image processing systems to determine if the vessel is using illegal drift nets in an area where these nets are prohibited. If so determined, the drone would then fire off a message via satellite to the coast guard to dispatch officers and arrest the vessel's crew.

Note

1. The Economist. (July 7, 2018). *Avast, Me Hearties: If Drones Ruled the Waves.* https://www.economist.com/the-world-if/2018/07/07/avast-me-hearties.

Figure 6.1 Unmanned Surface Vehicle That Would Carry a Drone (Roza Minasyan)

GPS Loggers on Albatrosses (SCP Technique # 7: Assist Natural Surveillance)

Wandering albatrosses forage over vast areas of water, flying up to 3,000 km and covering over 10 million sq km, in search of food. Their natural habit of flocking around fishing vessels in search for food makes them ideal spies against vessels engaged in illegal fishing. Taking advantage of the natural movements of the wandering albatross, ecologist Henri Weimerskirch and his team developed a way to detect the movements of fishing vessels. They designed a GPS logger that detects the radar signals from vessels nearby along with the positions of the birds that are fitted with lightweight tracking devices. The researchers conducted their pilot study on the Crozet Islands where they fitted 53 albatrosses with tracking devices (loggers). These devices were removed from the albatrosses once they returned to their nests. Not surprisingly, there was high correlation between the movements of the birds and the fishing vessels. Because so many albatrosses forage in the high seas, it is difficult to detect those that are wearing these logging devices, making them ideal spies against illegal fishing.[12]

WildLeaks (SCP Technique # 9: Utilize Place Managers)

WildLeaks is a digital platform used for collecting anonymous tips on illegal activities involving wildlife, such as hunting, poaching, logging, and fishing. The site provides a space for whistleblowing on the industry and allows

---

Box 6.3 Robotic Fish

Researchers at the Dynamical Systems Laboratory at the New York University Tandon School of Engineering have developed a 3D printed robotic fish. The purpose of this robotic fish is not only to study the behavior of real living fish in a laboratory setting, thus sparing the lives of real fish in experiments but also, importantly, to use these robots to drive schools of fish away from danger.[1] The robotic fish is programmed to move using two different approaches: independently of the real fish (thus, leading them into a controlled direction) and by tracking and mimicking the real fish movements. In laboratory experiments, the researchers were able to show that the fish accepted the robot into their school as they would a real fish, albeit some were wary of their robot counterpart.[2] This technological innovation can play an important role as a tool to detect and prevent illegal fishing in the future.

Notes

1. Scott, C. (June 5, 2018). *Saving Real Fish with 3D Printed Robotic Fish*. https://3dprint.com/215846/3d-printed-robotic-fish-real-fish/.
2. Yirka, B. (February 24, 2012). *Robot Fish Found Able to Lead Real Fish*. https://phys.org/news/2012-02-robot-fish-real-video.html.

---

individuals, who may not otherwise come forward for fear of retribution, to share information that would help catch hunters, traffickers, shippers, and buyers. Government officials can use the information posted on the site as evidence to trigger investigation and lead to subsequent prosecution. The following is an excerpt of a tip posted on February 16, 2015 (which was also accompanied by a picture not shown here), regarding a manta ray slaughter in the Philippines.

> I was recently in the barangay of Bungamar on the island of Bohol, Philippines and witnessed the slaughter of a very large Manta that was caught by fisherman. It was then systematically butchered by several in the barangay and "processed" and dried right there. It was my understanding that Mantas are protected and this was an illegal act. Is there anything that can be done?[13]

Local Initiatives

Empowering Artisanal Fishers to Take Charge of Their Fisheries and Lead the Fight against Illegal Fishing: The Case of Guinea (SCP Technique # 6: Extend Guardianship)

Guinea is one of the countries where illegal fishing is rampant. In 2000, Guinea initiated an effort where staff from the National Center for Fisheries Surveillance and Protection (CNSP) trained artisanal fishers fishing in local communities to take charge of their fishing areas and monitor the fishing activities there. This particular area, where the program took place, has had hundreds of illegal incursions by illegal fishing vessels. These incursions often resulted in collisions with the canoes and boats of artisanal fishers and the destruction and loss of fishing gear, boats, and even life. The program gave artisanal fishers various equipment, such as portable radios, GPS units, waterproof torches, and life jackets. These fishers patrolled the 12-mile coastal zone that was reserved exclusively for artisanal fishing. They were instructed to immediately contact the nearest CNSP station if they spotted a vessel fishing illegally in the zone. The CNSP sent out a patrol boat to further investigate the activities of these vessels and apprehend if the vessels were caught fishing illegally. The program, which only cost $20,000 to implement, saw significant success, with a 60 percent drop in illegal incursions by 2002. This targeted patrolling also allowed the CNSP to focus their efforts where they would achieve most promising results. Encouraged by the success of this program, other West African countries, namely Mauritania, Gabon, and the Congo, replicated this effort in their EEZs.[14]

Recruiting "Citizen Inspector" Volunteers to Lead the Fight against Illegal Fishing (SCP Technique # 9: Utilize Place Managers)

Visiting a Royal Air Force Museum in north London in 2013, and learning about the work of the Royal Observer Corps, uniformed volunteers who spent hours scanning the sky for German planes, gave Van der Werf, the founder of the Black Fish[15] organization, an idea. He established the Citizen Inspector Network that recruits "undercover" volunteer fisheries investigators. These investigators gather intelligence about illegal fishing activities at coasts, fish markets, and ports using a variety of techniques and tools, such as posing as tourists and engaging in conversations with fishers suspected of offloading illegally caught fish or using selfie-sticks to take surreptitious photos. The team comprises about 150 members, 85 percent of whom are women between the ages of 25 and 35, and its work is gaining strong momentum in Europe.[16]

Encouraging Responsible Fishing through the Sustainable Livelihoods Approach (SCP Technique # 24: Assist Compliance)

In an effort to address unsustainable fishing practices and reduce poverty in small-scale fishing communities, the UN FAO, in partnership with the Department for International Development of the United Kingdom of Great Britain and Northern Ireland and 25 countries in West and Central Africa, developed and implemented various Sustainable Fisheries Livelihoods Programs. These programs were designed to not only reconcile poverty reduction and encourage responsible fishing but also show the small-scale fishing communities that such approach to fishing is essential to ensuring long-term sustainable outcomes. Two examples—one from an inland fishery in Burkina Faso and one from a marine fishery in Guinea—are discussed in the following text.

In Burkina Faso, two management committees were set up to directly participate in the management of fish stocks in Lake Bagré and Lake Kompienga.[17] These management committees comprise territorial administrators, consular chambers, nongovernmental organizations, microfinance institutions, and representatives of traditional rulers and socioprofessional associations. The committee, which was assigned legal authority to approve and validate comanagement plans and rules of procedure, worked through different specialized commissions to address the specific needs of the fishing communities and protect their interests. By working directly with the members of the fishing communities and involving them in the decision-making process, these committees gained credibility and trust among the members of these communities and gained their cooperation and participation in the comanagement process, which involved such activities as surveillance in fishing camps, protection of fish habitats by creating fish reproduction zones, and protection of the water banks.[18]

In Guinea, community development committees were formed to encourage participation in fishery resource management. These committees, in collaboration with the fishing communities, engaged in participatory monitoring and surveillance, where the small-scale fishers were assigned the important role of not only active onlookers but also active decision makers in formulating and implementing fishery management regulations.[19]

Box 6.4 Timor Leste and the First-Ever Stop IUU Fishing Award

Timor Leste, with a population of 1.2 million, became a sovereign state in 2002. It is one of the least developed countries in the world, ranking 150th out of 177 countries on the Human Development Index. More than two-thirds of the population is employed in the fishing sector.

In 2014, the country was the first ever to win the Stop IUU Fishing award for its project *A Community-Based IUU Reporting System.* The award contest was launched by the International Monitoring, Control, and Surveillance Network, in an effort to encourage local communities to combat illegal fishing, and the entries were evaluated based on three assessment criteria: tangible impact in reducing IUU fishing, creativity and innovation, and feasibility and cost.

The project was designed to develop collaboration between the government agencies responsible for enforcing fisheries laws and artisanal fishers who are directly impacted by illegal fishing. To build this collaboration, the government of Timor Leste loaned GPS locator beacons, the SPOT© tracker, to artisanal fishers who were willing to collaborate. Using these trackers increased fishers' safety at sea. For examples, these fishers' vessels could be tracked by their families, and, in case of emergency, the vessels would receive immediate assistance. In return, these fishers were asked to use the device to report any illegal fishing activity that they encountered at sea. The device that was loaned to these fishers had two buttons: the "illegal fishing" button, which, when used, would send an emergency distress signal with the vessel's exact location, and this signal would be captured by the Maritime Police, who would then take appropriate measures to respond, and the "emergency" button, which the fishers used in case of emergency at sea. This collaboration not only strengthened the relationship between the artisanal fishers and the government but it also helped build the first-ever database in the state on illegal fishing hot spots and increased the efficiency of inspection and apprehension of illegal fishing vessels.[1]

Note

1. International Monitoring, Control and Surveillance Network. (2014). *Timor Leste Wins First-Ever STOP IUU Fishing Award.* http://gfetw.org/wp-content/uploads/2013/08/Press-Release_Feb-20_English.pdf.

Encouraging Compliance through Self-Regulation: The Role of the Rights-Based Management Programs (SCP Technique # 24: Assist Compliance)

Rights-based management programs take a unique approach to fisheries management. These programs are based on allocating fishing rights to legal fishermen and are designed to shift the management responsibilities of the fisheries from the governments to these fishers. Fishers are incentivized to maintain and restore the fish populations by following scientifically established catch limits. This ensures a healthy growth of the fish populations in the fishery and, eventually, leads to the increase of catch limits in the future. Evaluations of these types of programs have recorded not only moderate increases in catch shares but also up to 80 percent drops in ghost fishing (abandoning fishing gear at sea that kill marine species), dramatic drops in bycatch, and about 50 percent reduction of discards.

There are over 200 such programs in place in both developing and developed countries around the world. Some of these are territorial use rights for fishing (TURF) programs, where a group of fishers is given exclusive privileges to a fishing area or a fishery, and it becomes their responsibility to ensure that they not only self-regulate their fishing activities but also monitor and report possible illegal fishing in the area. The quota-based programs, which include community fishing quotas, individual transferrable quotas, and individual nontransferrable quotas, are based on assigning the allowed catch to participating fishers as shares, who are incentivized to maintain their catches within the scientifically established catch limits so that fish populations grow and the fishers benefit in the long run as a result.[20] The individual vessel quota system is a management method that allocates the total allowable catch among different fishing fleets, such as long-liners, trawlers, and purse seines. Each vessel group is then allocated a group quota that is to be shared among the participating fishing vessels.

Box 6.5 Case Study: The Spanish Galicia Goose Barnacle Cofradía System

Goose barnacles are a Spanish culinary delicacy. Cofradías, the traditional fishing guilds, harvest these crustaceans by using scraping devices to remove them from rocks. The increasing market demand for goose barnacle, coupled with the cofradías' inability to prevent outsiders from fishing within their territories, led to the near collapse of this fishery in the late 1980s. The initial attempt to regulate the fishing efforts with temporal closures proved to be unsuccessful. In 1992, the Galician government established the territorial use of rights for fishing (TURF) program for the goose barnacle, giving cofradías exclusive fishing access, as well as assigning them the management responsibility. The cofradías were assigned TURF boundaries based on their traditional fishing ground and were, therefore, responsible for managing these fishing grounds. In order to retain their fishing rights, cofradías had to develop yearly management plans that included information on fishing locations, designated control points, points of sale, total expected catch in a given year, and daily individual catch limits, among others. Importantly, their catches were sold through a designated market known as Lonxa. To sell the goose barnacles at the Lonxa market, dealers had to show inspectors the invoice issued to them by the cofradías who had sold them the crustaceans, thus ensuring the legality of the source. This approach led to the recovery of goose barnacle stocks, which remain stable and healthy today.[1]

Note

1. Fishery Solutions Center. (2018). *Design Stories: Spanish Galicia Goose Barnacle Cofradía System.* http://fisherysolutionscenter.edf.org/design-stories/spanish-galicia-goose-barnacle-cofradía-system.

Other Notable Initiatives

Refusing Vessel Insurance to Illegal Fishing Vessels (SCP Technique # 2: Control Access to Facilities)

Oceana, one of the major nonprofit organizations in the world that works on ocean conservation, recently launched an initiative that targets vessel insurance companies. The organization developed guidelines to help these companies assess, control, and mitigate the risks of insuring vessels and companies linked to illegal fishing activities before insuring them. Oceana recommended against insuring illegal fishing vessels because of the following risks associated with it:

- Increased likelihood of claims
- Increased possibility of association with other crimes
- Increased exposure to fraud
- Increased exposure to legal liabilities
- Increased possibility of losses due to reputation damage

The organization also developed detailed guidelines (shown in Table 6.1) pertaining to vessel checks, disclosures, and other operational procedures to help these insurance companies in their decision-making process.

Table 6.1 Oceana Guidelines for Insurance Companies

Vessel Checks
• Name—ascertain if present on IUU vessel list
• Number of name changes; more = worse
• Flag—flag of convenience of noncompliant flag = increased risk
• Number of flag changes; more = worse
• IMO number—should have one if vessel over 100 GT
• Classification society, if appropriate
• Last survey—all recommendations demonstrably carried out
• Records evidencing regular maintenance
• Crew nationalities and experience
• Owner's claims record
Disclosures
• Any previous convictions
• Any fines or prohibitions
• Any previous refusal of insurance
Indicative Operational Factors
• Operational area—whether known to have IUU fishing activity
• Involvement with high-risk species, as defined by the *Petrossian List*
• Use or otherwise of vessel tracking systems (e.g., VMS or AIS)
• Absence of fishing license/authorization
• Multiple claims for lost gear

Establishing Marine Reserves (SCP Technique # 1: Harden Target)

Many conservation groups and marine scientists have argued that establishing marine reserves, often known as "no-take zones" or "marine sanctuaries," is an effective way to deal with overfishing. These areas bar fishing and restrict other human activities. This, in theory, would lead to the recovery of the depleted fish populations within the reserve and the consequent "spillover effect" that would also replenish nearby fishing grounds. Such results, however, would only be possible if the reserve is strictly protected from fishing for several years, which would depend on the reproductive capacity of the species protected in the reserve. Nevertheless, a study that reviewed 73 reserves around the world found that the overall biomass of the fish inside these reserves—their density, size, and diversity—was significantly higher than those in the control sites outside the reserve.[21]

Reducing Fishing Subsidies (SCP Technique # 5: Control Tools)

Fishing subsidies are various forms of government assistance programs for the fishing industry. These programs may range from direct payments to the fishing industry to indirect aid, such as building fish handling facilities at harbors or supporting research on new fishing gear. The Asia-Pacific Economic Cooperation divides fishing subsidies into six types:

1. Direct assistance to fishers and fishery workers (e.g., unemployment insurance)
2. Lending support programs (e.g., low interest rates on loans to purchase fishing vessels or equipment)
3. Tax preference and insurance support programs (e.g., government-funded vessel insurance)
4. Capital and infrastructure support programs (e.g., paying access fees to the fishing waters of other nations and building harbor facilities and other port infrastructure)
5. Marketing and price support programs (e.g., advertising and promoting seafood products)
6. Fisheries management (e.g., vessel buyback programs)[22]

While some of these subsidy programs can be very beneficial, especially for small-scale fisheries, many have argued that some of these fishing subsidies contribute to unsustainable fishing. For example, many environmental organizations, including Oceana, believe that *lending support programs* are a leading contributing factor to the excessive number of the world's fishing fleets, which is about 250 percent larger than the world's fish stocks can withstand.[23] European *capital and infrastructure support programs*, mainly those that include paying access fees to allow European vessels to fish in West African EEZs, have also been criticized because such programs lead to the depletion of West African fish stocks targeted by the European vessels.

Recognizing that fisheries subsidies are a leading contributing factor to economic losses in the fisheries; can disrupt food security, employment opportunities, and livelihoods of developing countries' coastal communities; and can also encourage IUU fishing, the 27 members of the World Trade Organization (WTO) have committed to reforming subsidy programs.[24] The WTO continues to engage its member states into a dialogue on how subsidies reduction can be accomplished and what this would mean for fisheries management and conservation worldwide.

MONITORING AND CONTROL THROUGH TRADE-RELATED MEASURES (SCP TECHNIQUE # 21: SET RULES)

The Convention on the International Trade in Endangered Species

Several international measures have been undertaken in the past five decades to address the issue of effective sustainability and conservation of biodiversity. While pertaining to biodiversity and conservation, these measures have strong implications for fisheries as well.[25] Among the most notable of these measures is the 1948 International Union for Conservation of Nature and Natural Resources (IUCN) Red List of Endangered Species assessment. The IUCN Red List is established by scientifically assessing the status of both terrestrial and marine species and assigning status to these species based on their vulnerability to become extinct. One of the major contributions of IUCN was the initiation of the establishment of an international convention to govern the trade in animal species and their products, namely the drafting of the Convention on International Trade in Endangered Species of Wild Fauna and Flora (CITES).[26]

The first draft of the CITES was produced in 1964, and it went into effect in 1975 after 21 nations signed the convention in Washington, D.C.[27] To date, CITES has 183 member states, and it governs the international trade in 5,800 animal and 30,000 plant species in an effort to prevent the overexploitation of these species. The CITES signatory countries, known as the Parties, meet every two years at the Conference of the Parties (COP) to discuss the state of select species, consider efforts that have been implemented to safeguard these species, and propose solutions and recommendations. For example, during the 2005 COP, the bluefin tuna was among the species highlighted as needing much more protection than it was receiving.

Among the most important contributions of CITES is the listing of the species in appendices (I, II, and III). The placement of the species into these appendices depends on how threatened these plant and animal species are by legal commercial trade. There are several factors that Parties consider when determining the status of a certain species or when reviewing proposals to consider listing new ones or removing those in these appendices. These include the following:

1. Is the species being traded?
2. What is the population size of the species, and is it at a risk of decline?
3. What is the quality of the habitat the species uses?
4. What is the geographic distribution of the species?
5. How many subspecies does the species have, and what is the population size of these subspecies?
6. What is the reproductive potential of the species?[28]

Species that are threatened with extinction primarily due to commercial trade are listed under Appendix I, prohibiting all trade. Under some exceptional circumstances, such as for scientific exchange, breeding, or educational programs, these species can be traded, but the trade must be accompanied by both an import and export permit. Some marine species listed under Appendix I include the totoaba (*Totoaba macdonaldi*), the bowhead whale (*Balaena mysticetus*), the Guadalupe fur seal (*Arctocephalus townsendi*), and the Atlantic sturgeon (*Acipenser sturio*).

Appendix II species, among which are the great white shark (*Carcharodon carcharias*), the European eel (*Anguilla anguilla*), and the giant wrasse (*Cheilinus undulatus*), can be commercially traded, but only if the origin country procures an export permit and produces proof that the trade will not be detrimental to the survival of the species in the wild. In other words, these species are not immediately threatened with extinction, but they need to be closely regulated as they can potentially become threatened. In certain cases, CITES handles Appendix I species as Appendix II, such as in the cases when the species have been bred in captivity or artificially propagated. Importantly, in order to import either Appendix I (under exceptional circumstances) or Appendix II specimen taken from the high seas outside of the countries' EEZs, the trade must be accompanied by a "from-the-sea certificate."[29] Similarly, some Appendix II-listed species can be listed as Appendix I if they come from certain geographic regions or countries. For example, the African clawless otter (*Aonyx capensis microdon*) from Cameroon and Nigeria is listed as an Appendix I species; all other *Aonux capensis microdon* otters are listed in Appendix II.

Lastly, species listed in Appendix III can be traded when accompanied by an export permit and a certificate of origin. Providing proof that the trade will not be detrimental to their survival is not mandatory in this case. When party to CITES, countries are required to implement local legislation governing the trade in the species and providing penalties for violations. In the United States, for example, the provisions of CITES are implemented through the Endangered Species Act, the Lacey Act, and the Pelly Amendment of the Fishermen's Protective Act.[30]

Apart from CITES, there are a number of other international instruments that deal with biodiversity and conservation. Among these are the Convention on the Conservation of Migratory Species of Wild Animals (Bonn Convention, 1983), the Convention on the Conservation of Biological Diversity (1992), the International Convention for the Prevention of Pollution from Ships (1983), and the Global Plan of Action for the Protection of the Marine Environment from Land-Based Activities (1995). These conventions provide a useful framework to address issues related to sustainability, preservation of habitats, and preservation of biological diversity.[31]

RFMOs and Their Role

RFMOs are intergovernmental organizations that have the mission of establishing and facilitating fisheries conservation and management measures. RFMOs are composed of members from different fishing nations: those that are in the region, that is, coastal states and the states that have interest in the fisheries managed by the particular RFMO.

RFMOs coordinate access to mutually shared resources that are beyond the countries' EEZs. This access is generally determined based on the assessments provided by committees of experts employed by the given RFMO, the health of the fishing stocks, and the potential impacts of commercial harvesting on the viability and sustainability of the fisheries. These assessments then lead to a set of recommendations brought forward to the organization, which either adopts or rejects these recommendations. If accepted, the RFMO adopts these recommendations as regulatory and operational guidelines and communicates these new sets of rules to member states to follow until another set of recommendations is proposed.[32]

While RFMOs engage in many functions, their roles can be classified into three major types: (i) those that conduct scientific research, (ii) those that perform advisory and regional coordination roles, and (iii) those that are engaged in management and governance of fishery resources.[33] The central role of the latter is to develop measures designed to address the problem of IUU fishing within their management areas. To achieve this goal, RFMOs:

1. *Set and allocate quotas* for the fish stocks under their management, and *enforce these quotas* through various *control, monitoring, and surveillance activities.* These activities include, but are not limited to, establishing port control measures, such as granting authorization to transship or land fish at a member state's port or reporting, in a timely manner, the entry and exit of the vessel from the port of the state that is party to the RFMO. To govern these transshipment operations, RFMOs require that (a) vessels carry observers in accordance with observer programs established by the relevant RFMO; (b) port inspectors of the member states inspect transshipment declaration and

verify other relevant documents; and (c) member states report, on an annual basis, the quantities of species transshipped via their ports and provide a list of vessels that conducted these transshipments.[34]

2. *Monitor the conservation and protection of fish stocks on the high seas* and those migrating through the waters of more than a single state. To do this, the RFMOs take measures to ensure that only authorized vessels are fishing in their area of competence. For example, they establish and monitor various observer programs and trade documentation schemes, as well as set regulations that pertain to satellite-based VMS that allow the identification and tracking of the vessel's position and movements.
3. *Monitor the conservation of stocks that fall within the EEZs of the states that are party to the RFMO.* While the member states have the authority to grant access to their fishing grounds that fall within the RFMO's convention area, these states are, nevertheless, mandated by the RFMO to issue such authorization only if they have taken effective control measures pertaining to licensing and fishing regulations that cover a range of conditions related to specific fishing areas, species caught, time periods for which fishing is authorized, reporting of catch data, and operating a VMS.
4. Design measures to conserve all species associated or affected by their fisheries, such as turtles, dolphins, seabirds, sharks, and nontarget fish. These may include *establishing bycatch mitigation measures* or prohibiting the use of certain gear to catch a target species.
5. *Create and maintain "blacklists"* of vessels engaged in illegal fishing. This is done in several steps. Initially, an inspection authority of a member state sights a vessel due to a violation. The member state then informs the relevant RFMO about the sighting and provides relevant information and supporting evidence of the violation. The RFMO provides all member and cooperating nonmember states, as well as the flag state of the vessel (where the flag state is not a member of the given RFMO), information and the supporting documents about the sighting. This procedure serves two purposes: to obtain feedback from the member and cooperating nonmembers about the provisional listing of the vessel and to give the flag state the opportunity to provide evidence that the vessel in question has not violated the RFMO's or the given state's fisheries regulations. After the examination of the submitted evidence, as well as the feedback, the RFMO decides whether to blacklist the vessel or not. If the RFMO makes a decision to blacklist the vessel, it is placed on a provisional publicly available list that includes information about the vessel and its activities that were deemed illegal. The vessel can be removed from the provisional list if the flag state is able to demonstrate that it had not engaged in illegal fishing activities or, since being placed on the provisional list, it has taken effective measures in response to the violation in question.

RFMOs may focus on certain species of fish. For example, the International Convention for the Conservation of Atlantic Tunas (ICCAT) focuses on the

Figure 6.2 ICCAT Convention Area

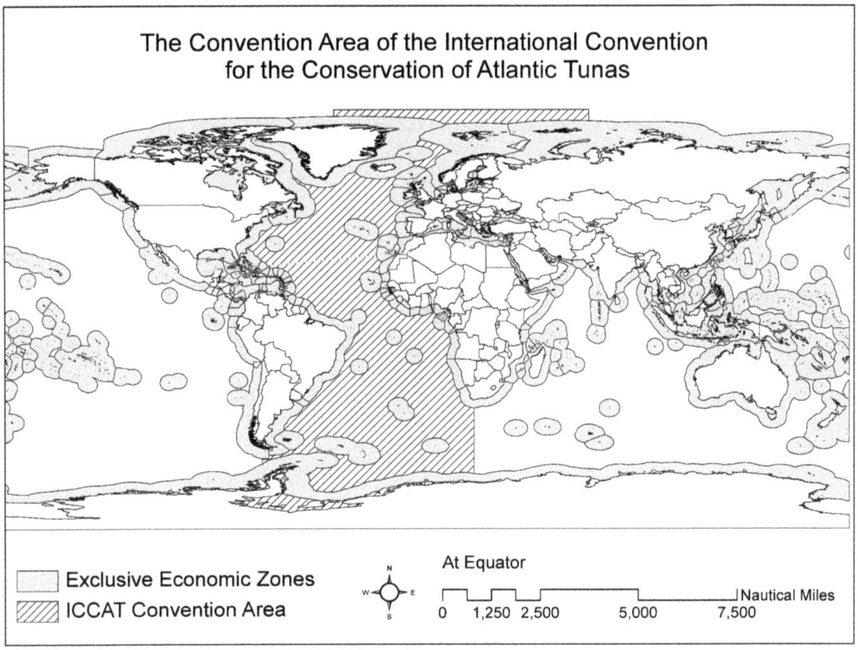

conservation of tuna within its convention area that spans the entire Atlantic Ocean (see Figure 6.2).

Other RFMOs may have a wider agenda related to all living marine resources within their convention area, unless specified otherwise. For example, Figure 6.3 shows several RFMOs that have management authorities within specified segments of the Atlantic Ocean. The Northwest Atlantic Fisheries Organization (NAFO) manages all fish stocks in the Northwest Atlantic, except for salmon, tuna, whales, and shellfish. Meanwhile, the South East Atlantic Fisheries Organization (SEAFO) deals with the management of all living marine resources within its convention area. Often, the convention areas of RFMOs overlap, such as the areas of the RFMOs that manage parts of the Atlantic Ocean shown in Figure 6.3. In this case, it becomes even more important that these RFMOs cooperate in order to synchronize their efforts.

The EU-IUU Regulation

The European Union is the world's largest importer of seafood products. As such, it carries a profound responsibility for establishing and enforcing regulations that deal with illegal fishing. One such effort is the EU-IUU Regulation.

Figure 6.3 Convention Areas of Select RFMOs in the Atlantic Ocean

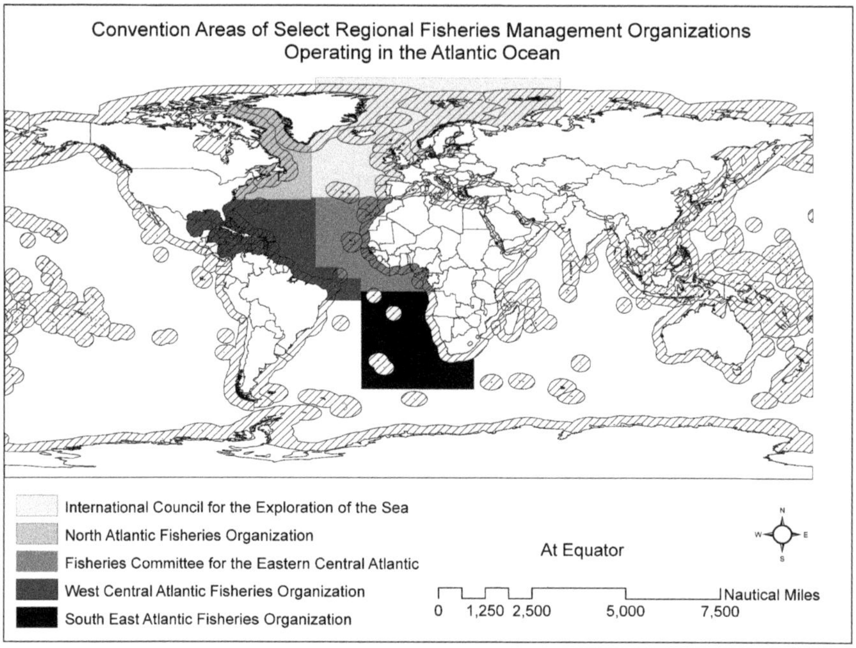

The regulation not only aims at deterring and eliminating the trade in illegally caught fish within the European Union but it also deals with countries that export fish into the European Union. This is carried out through three operational strategies:

1. The only fishery products that can be sold in the EU markets are those that carry *catch certificates* showing that the fishers operated in compliance with fisheries laws and conservation measures when removing the fish. This requirement primarily pertains to EU member states and ensures that the fish both entering and exiting the EU markets are not only traceable but they also come from verifiable and legal sources.
2. The European Union imposes *substantial penalties* if the *member state* national is caught to have engaged in illegal fishing. For the first offense, the fishing vessel receives a penalty that is five times the value of the fish caught illegally. For any subsequent repeat offense that takes place within a five-year period, the vessel's penalty amounts to eight times the value of the illegally caught fish. This is done in an effort to increase the cost and reduce the benefit of engaging in illegal fishing. Lastly, if EU nationals support or engage in illegal fishing anywhere in the world, under any country's flag, they will receive penalties that are proportionate to the economic value of the catch.

In an effort to regulate the EU member states' operations, the EU-IUU Regulation also requires that these member states designate ports for third-country vessels to land their catch and that these states inspect at least 5 percent of all landings at their ports.[35]

3. The last operational strategy pertains to *non-EU countries* that import fish into the European Union. This strategy involves a *carding process* and includes assigning these non-EU countries green (good), yellow (warning), and red (failed) cards. The EU issues these cards based on assessments of the adequacy of these non-EU countries' fisheries management and conservation measures in dealing with illegal fishing within their waters and by their vessels. A red card, for example, means that the European Union will impose a temporary trade ban on the fishery products from that country until the country demonstrates that it has undertaken significant measures to deal with illegal fishing. In essence, this third-country carding strategy assesses these non-EU countries' monitoring, control, and surveillance efforts, and this assessment is based on three criteria:
   a. Does the country's legal framework comply with international fisheries management and conservation requirements, such as those pertaining to vessel registration and monitoring, as well as inspection, enforcement, and sanctions?
   b. Has the country ratified international instruments, and does it participate in multilateral cooperation, including being a member of the relative RFMO and complying with the RFMO's conservation and management measures pertaining to observers, reporting, and maintenance of lists of authorized vessels?
   c. Does the country implement appropriate fisheries and conservation measures, such as establishing accurate licensing systems and maintaining updated lists of authorized vessels to ensure that its vessels are fishing legally both within and beyond its EEZ?[36]

MONITORING AND CONTROL THROUGH MARKET-RELATED MEASURES (SCP TECHNIQUE # 23: ALERT CONSCIENCE)

A number of provisions within the IPOA-IUU have outlined measures referred to as Internationally Agreed Market-Related Measures. These measures primarily outline the procedures to be followed by countries when importing and exporting fish and fish products. For examples, IPOA-IUU recommends countries to establish traceability of fish to make it possible to follow it from catch to production, processing, and distribution. That is, the fish will be considered traceable if a system can be developed to track the fish "from the boat, sea, or farm to the table or fork,"[37] such as the examples discussed earlier.

Eco-labeling of fish and fish products is another example of a market-related measure and is related to the sustainability of fish stocks in general. These are "seals of approval" given to products that are determined to have a reduced impact on the environment. If a product has an eco-label, it is assumed that it has been harvested in compliance with sustainable fisheries management measures. In a way, it is a measure that suggests that, when informed about the product, consumers will more likely make the eco-friendly choices. There are three categories of eco-labeling, and these include those established by (i) individual companies and based on their product standards, (ii) industry associations, and (iii) private initiators.[38] The Ecolabel Index, a global directory of eco-labels, maintains a database of over 400 eco-labels from 199 countries and 25 industry sectors.[39] This index has a list of 27 eco-labels pertaining specifically to fish and fisheries. Table 6.2 provides a list and a brief description of these 27 eco-labels.

Table 6.2 A List of Eco-Labels on Fish and Fisheries

Eco-Label Name	Purpose
Australian Certified Organic	An ACO-labeled product is certified to be in compliance with Australian product standards, and it can be traced back to its origin.
B Corporation	Certified B Corporations are those that operate in compliance with standards of social and environmental performance, transparency, and accountability.
Best Aquaculture Practices	BAP certification standards confirm that the seafood products come from BAP-certified facilities (e.g., shrimp farms and seafood processing plants) that operate in compliance with the BAP standards, which deal with environmental, social, food safety, and traceability issues.
BIO Hellas	An approved inspection and certification body by the Bulgarian and Moldavian Ministries of Agriculture and Food, BIO Hellas provides certificates that attest to the organic nature of the products.
Biokreis	This eco-label helps farmers grow organically.
Bio Suisse	Label indicates products produced in Switzerland that are totally organic.
Carbon Neutral Product Certification	This eco-label ensures that in the process of the production and distribution of the product, no CO_2 emissions were used in raw materials, fuel, and electricity.
Certified Green Restaurant®	Restaurants receive this eco-label if they have demonstrated commitment to sustainability.
Dolphin Safe/ Dolphin Friendly	Tuna companies receive this eco-label if the Earth Island Institute, which monitors their operations, ensures that these companies caught tuna using methods that are not harmful to dolphins, and protect the marine ecosystem.

Table 6.2 (Continued)

Eco-Label Name	Purpose
Earthsure	The Earthsure eco-labeled products disclose information on the environmental impact the creation of the product has.
Eco-Leaf	Similar to Earthsure, this is an environmental product declaration eco-label that traces the life cycle of a product to determine its impact on the environment.
EcoMark Africa	Currently in its development stage, this eco-label specifically pertains to products from the African continent.
Environmental Product Declaration	This eco-label is also a label that traces the life cycle environmental impact of products.
Fair for Life	This eco-label promotes social accountability and fair trade in agricultural, manufacturing, and trading operations.
Fair Trade Certified	This eco-label certifies that the workers were paid better prices and wages, work in safe conditions, and the product has been manufactured with the protection of the environment in mind.
Green Tick	Products under this eco-label have undergone life cycle audits.
LFP Certified	Local Food Plus-certified products come from local sustainable food sources.
LowCO_2 Certification	Organizations with this certification have undergone greenhouse gas audits.
Marine Stewardship Council	MSC-certified fish are those that have met the three requirements: (1) these are sustainable fish stocks; (2) their harvesting did not result in major environmental impacts (such as by-catch of other species); and (3) the fisheries where these are harvested from are well managed and met all local, national, and international laws.
Milieukeur: the Dutch environmental quality label	This eco-label is a Dutch environmental quality label for products and services.
Naturland e.V.	This eco-label promotes organic farming.
OceanWise	This eco-label appears on menus of restaurants that have worked closely with OceanWise to ensure that they make ocean-friendly buying decisions.
Ø-label: Norway	This is a Norwegian eco-label that attests to the organic nature of the products, as well as has undergone Debio's inspection, which traces their origin from farm to table.
SeaChoice	SeaChoice, which is a sustainable seafood program in Canada, endorsed products if they come from fisheries that have undergone evaluations based on the stock status, species inherent vulnerability, and habitat/ecosystem impacts.

(Continued)

Table 6.2 (Continued)

Eco-Label Name	Purpose
SEE What You Are Buying Into	This is an eco-labeling for companies that are open about their social, environmental, and ethical policies and practices.
Soil Association Organic Standard	This association provides organic certification to farmers, food processors, packers, retailers, and so on.
Sourcemap	The Sourcemap eco-label, when scanned, directs consumers to their website (sourcemap.com) that provides information on the product's supply chain that can be traced, as well as its environmental footprint and social impact.

CONCLUDING REMARKS

International, regional, and national actors have proposed a plethora of strategies to deal with the problem of illegal fishing. In an effort to show the readers the breadth of the efforts undertaken worldwide (but by no means all encompassing), this chapter discussed some of the most important, innovative, and promising approaches proposed and implemented to date that are designed to *increase the effort, increase the risk*, *reduce the rewards,* and *remove excuses* of engaging in illegal fishing. Some of these approaches take advantage of available technological tools to trace the "journey" of the fish from "bait to plate" or engage in local, regional, or international surveillance of illegal fishing vessels, thus *reducing the rewards* of such activities. Others pull together global intelligence and human resources to *increase the effort* and *risks* of continued involvement in illegal fishing. The measures that are designed encourage compliance and discourage imitation range from strict regulatory mechanisms to advisory strategies built upon the provisions of various United Nations' legally and nonlegally binding instruments. Unique to these strategies are self-regulatory incentive-based programs that are designed to *remove excuses* by "assisting compliance."

Chapter 7

Can Illegal Fishing Be Stopped?

THINKING ABOUT ILLEGAL FISHING THROUGH THE LENS OF ENVIRONMENTAL CRIMINOLOGY

In this book, we emphasized the importance of studying illegal fishing, essentially a crime, through a criminological lens. We specifically focused on two perspectives, rational choice and situational crime prevention, to build an understanding about the global illegal fishing patterns and highlight prevention strategies.

The rational choice perspective posits that a "motivated offender" will consider the risks, efforts, and rewards and decide upon this calculus to engage in criminal acts accordingly. Consequently, increasing the presence of patrol surveillance, as well as strengthening the monitoring, control, and surveillance (MCS) capacity of the fisheries management organizations, will deter illegal fishers. The presence of observers onboard fishing vessels will "increase the risk" of being caught fishing illegally, as these observers will closely monitor fishers' activities to ensure that they neither catch above the allowable quota nor catch prohibited species. Vessel monitoring systems (VMS), in turn, will allow the control centers to closely observe the activities of fishing vessels, and detect possible violations. Increasing inspections at known ports of convenience (POC) will mean "increasing the efforts" the potential offenders will have to make to get their illegally caught fish on land, as they would need to look for new locations or travel longer distances, which may not be desirable or possible. This will also mean reducing the opportunities for the illegal fishing vessel operators to get their catch into the markets on time. Therefore, it is expected that a combination of these MCS measures will increase the potential offenders' perception of risk of being caught, and reduce the expected rewards.

The 25 techniques of situational crime prevention can be applied to focus specifically on these elements. Designed around *increasing the risk, increasing the effort*, and *reducing the reward* (as well as *removing excuses* and *reducing provocations*), the situational crime prevention strategies provide a useful framework to understand how offenders act spatially. The decision to engage in illegal fishing would depend not only on the availability of the resources sought but also, due to weak surveillance, and the availability of unscreened "exits." Many situational crime prevention techniques are, in essence, already being used around the world to deal with the problem of illegal fishing (as shown in Chapter 6). For example, mandating fishing licenses and VMS is an example of "controlling access to facilities"; inspecting catch at ports is a "screening the exits" measure; requiring vessel registrations means "reducing anonymity"; and utilizing observers onboard fishing vessels is one way of "strengthening formal surveillance." It is worth noting that these MCS measures that are, in fact, situational were the strongest predictors of illegal fishing on the basis of the analysis of the data in Chapter 5. Additionally, among the successful and promising strategies that have been implemented to date to decrease the occurrence of illegal fishing are those that encourage artisanal fishers to take charge, which is an example of "extending guardianship"; recruit citizen inspectors to fight against illegal fishing, which is one measure of "utilizing place managers"; use blockchain technology or eco-labels to track the movements of the harvested species, which is a means of "disrupting markets" and "alerting conscience"; and encourage self-regulation to "assist compliance." The theoretical guidance from both rational choice and situational crime prevention perspectives proves important in understanding the decision to engage in illegal fishing.

ANALYZING ILLEGAL FISHING FROM THE ENVIRONMENTAL CRIMINOLOGY AND CRIME SCIENCE PERSPECTIVE

We are optimistic that illegal fishing can be stopped. This is possible if (a) more *data-driven analyses* are conducted to better understand the crime and (b) the prevention strategies that are developed and implemented utilize the toolbox of the techniques of *situational crime prevention*. Here, we will outline some analytical strategies that have been used by crime analysts to deal with everyday crimes and demonstrate ways that these can be or have been applied to address illegal fishing.[1]

Environmental criminology emphasizes the importance of being crime specific, paying attention to environment-person interaction, and analyzing concentrations in time, space, victim, and offender. Being crime specific in the case of illegal fishing is especially important. In Chapter 2, we discussed what constitutes *illegal* fishing, and using the UN FAO definition, we outlined five different scenarios that can be considered as such, including the following:

- Using prohibited gear or methods
- Operating in closed areas or during closed seasons
- Operating with a fake license, without a license or vessel registration
- Operating without a VMS
- Fishing beyond allocated catch limits or taking prohibited fish

An analysis of using prohibited gear or methods, that is, strategies that pertain to tools or the *modus operandi* of illegal fishing, would require an in-depth understanding of the steps involved in acquiring, using, and disposing of these tools; therefore, using *crime script analysis* would be one useful analytical strategy to use. This analytical strategy would also be suitable when studying the patterns of operating with a fake license or without a VMS.

To better understand the illegal activity that involves operating in closed areas or during closed seasons, a *spatiotemporal analysis* of the patterns and activities would be most helpful. To understand why illegal fishers take prohibited fish or catch fish beyond their allocated catch limits, one can apply the CRAVED[2] or CAPTURED[3] analytical models.

Additionally, the *risky facilities framework* can be applied to explore the characteristics of the ports most likely used by illegal fishers to off-load their catch, while *geographic profiling* can be used to understand the offender characteristics, networks, and patterns of interaction. In the following section, we briefly describe each of the analytical strategies outlined earlier and provide examples of how these strategies can be or have been used to analyze the problem and devise evidence-based prevention strategies.

What Is Crime Script Analysis?

Crime script analysis (CSA) is an analytical tool that allows the crime analyst to "organize and systemize knowledge"[4] about the process involved in carrying out a crime. This analytical strategy is premised on the assumption that in order to carry out a crime, the potential offender makes sequential decisions in stages that are largely interdependent. Therefore, the analyst focuses on understanding the activities that are involved in each of these stages and designing potential interventions to intercept the process, thus preventing the criminal from furthering the activity.

The original crime script involves seven sequential stages: *preparation, entry, precondition, instrumental initiation, doing, exiting*, and *aftermath*. To understand the crime of trafficking of wildlife, criminologists have also proposed a modified version of the script that comprises five stages: *poaching, processing, transporting, trading*, and *removing*.[5]

How Has CSA Been Used to Analyze Illegal Fishing?

Criminologists have examined the process of illegal fishing "from bait to plate" by building two sequential scripts. The first script (Stage 1) described the seven stages involved in the removal of fish and its landing at ports. Their analyses showed that the *preparation* stage, for example, involved obtaining coast guard training certificates; applying for license, permits, and vessel registration through legal or illegal means; purchasing vessel insurance; hiring crew; and setting navigation plans.[6] The second script (Stage 2) refers to the processing and distribution of the illegally landed fish at port and pertains more to the seafood fraud aspect of the crime. They found that the *preparation* stage involved obtaining fraudulent seafood import licenses, locating seafood warehouses nearby the landing port and obtaining a distribution permit, obscuring the certificate of origin of import, and preordering equipment, such as bench weight scales and storage (e.g., refrigeration/ice holdings). Based on their findings, the researchers proposed different intervention strategies that can be implemented at each of these "anchor points" to deal with the problem. For example, the researchers recommended that at the license application stage, the authorities consider denying the license application on the basis of past involvement in illegal fishing or association with a blacklisted vessel.

Interpol distinguishes three major phases involved in illegal trafficking of fish and detailed the types of crimes that were likely to occur at each stage.[7] The *preparation phase* involves any or all of the following activities: document fraud, corruption, tax evasion, and money laundering. The *catch phase* includes disobedience of an order to stop and forced labor, while at the *sale phase*, illegal fishers may be involved in the illicit trade and food fraud activities.

The CSA can be conducted by using either of these scripting strategies. The important point is that the process of illegal fishing should be broken down into sequential stages so that once the analyst has a clear understanding about how the act is carried out, he or she can propose prevention strategies that would disrupt the progression of this illegal activity.

What Is Spatiotemporal Analysis?

Spatiotemporal analysis refers to the set of tools used to understand the spatial and temporal distributions of crime. If illegal fishing concentrated in time and space, then such analyses are vital in understanding spatial and temporal patterns of this activity. Canadian environmental criminologists Pat and Paul Brantingham have devised the crime pattern theory to explain the geographic distribution of everyday crime.[8] They suggest that the decision to offend is largely influenced by the *geometry of crime*—that is, the environmental backdrop that provides the opportunities for crime. One of the central concepts of this theory is the idea of *activity spaces*. They suggest that crime is likely to happen when the activity space of the potential victim intersects with the activity space of the motivated offender. Note, these activity spaces are not randomly dispersed but are likely to be concentrated in time and space. Therefore, using the spatiotemporal analysis to understand illegal fishing will be useful in deciding where and when targeted intervention strategies can be implemented and patrol surveillance can be deployed to maximize their effectiveness.

How Has Spatiotemporal Analysis Been Used to Analyze Illegal Fishing?

In a recent spatial analysis of illegal fishing activities in the exclusive economic zones of 23 Western African countries, criminologists have discovered that this activity is not randomly dispersed but rather it is highly concentrated in space. The study found that these "hot spots" were in locations where there was a higher likelihood of finding CRAVED fish (discussed in more detail later) as well as those that are highly commercial, abundant, and valuable and are in close proximity to known POCs. If an area had both CRAVED and highly commercial species, this increased the likelihood of illegal fishing in that area by 61 percent. If the area had abundant and valuable species, it had a 26 percent chance of having illegal vessels fishing there. Lastly, the areas that were in close proximity to known POCs were 3.5 times more likely to have illegal fishing vessels than those that were farther away.[9]

In another recent study, researchers examined the temporal patterns of illegal fishing of red snapper in the Gulf of Mexico. They discovered that this activity is temporarily concentrated. Their findings revealed that illegal fishing clustered around three moon phases that were associated with the heights of tides and included the periods where wave heights were the lowest. While seasonality played a role, the morning hours were the best time for illegal fishing.[10]

What Are the CRAVED and CAPTURED Models?

In an effort to understand the attributes of "suitable targets"—that is, why some products are more frequently stolen than others—criminologists devised the *VIVA* (value, inertia, visibility, accessibility) acronym in 1979.[11] Most frequently stolen items were those high in value, easy to move, more common, and easily accessible. Two decades later, the CRAVED model was proposed to understand the characteristics of *hot products*, which are those that are easily concealable, removable, available, valuable, enjoyable, and disposable.[12] The expanded variations of this model include the CRAAVED model that separates *available* into *abundant* and *accessible* components[13] to understand the illegal poaching of wildlife, including illegal fishing,[14] and explain wildlife trafficking variations between open-air illicit markets,[15] and the CAPTURED[16] model that helps understand the characteristics of stolen wildlife beyond the initial poaching stage by focusing on the entire market continuum of the stolen wildlife from the point it is removed from the wild to the point of disposal. The CAPTURED model takes into account several factors, such as whether the wildlife needs to be processed (cut, altered); its longevity and perishability, which will affect its *concealability, usability, desirability*, and *transferability*; and whether the wildlife product is temporarily or permanently removed from the market (e.g., whether it is consumed, such as caviar, or resold, such as elephant ivory or tiger skins).[17] As such, this latter model is more suitable for studying the movement of wildlife through illegal wildlife markets.

How Has the CRAVED Model Been Used to Analyze Illegal Fishing, and How Can the CAPTURED Model Be Used to Expand on This Analysis?

In Chapter 2, we briefly discussed two studies that examined the characteristics of fish and crustacean species that were more sought after by illegal fishers.

These studies not only identified 58 high-risk fish and 10 high-risk crustacean species but they also applied the CRAVED theft model to understand the characteristics of these species. The 58 high-risk fish shared the following common characteristics: they were sold through more POCs, were caught with longline vessels, and were more abundant and accessible to known illegal fishing countries, more valuable (larger) species, more often found in recipes (i.e., enjoyable), and more easily disposable (i.e., highly commercial).[18] The 10 crustacean species (namely, lobsters and crabs) were targeted by illegal fishers because they were more abundant, valuable, and enjoyable.[19] These studies recommended that focusing on known POCs, monitoring longline vessels, educating consumers about these CRAVED species, and exerting pressure on known illegal fishing countries would be one way to stop illegal fishing of these species. These studies primarily focused on understanding the illegal removal of fish. In order to explore the theft continuum from "bait to plate," the CAPTURED model could be a more holistic analytical strategy.

What Is Risky Facilities Framework?

This framework helps analysts understand the characteristics of facilities of the same type (such as ports) that make them likely targets for carrying out illegal activities. Criminologists have argued that similar to geographic, temporal, and target concentrations, one is likely to see concentrations of crime within the homogeneous sets of establishments. Such important characteristics as the actual location of the facilities (e.g., whether it is located within a high-crime area), their design and layout, whether they draw in large number of offenders, their size, and their poor management are some of the characteristics that can explain the variability in risk of those facilities being targeted by motivated offenders.[20]

How Has the Risky Facilities Framework Been Used to Analyze Illegal Fishing?

Researchers have used the risky facilities framework to study the characteristics of POCs where illegally caught fish were off-loaded. They found that POCs that were most attractive were those that made it easier to conceal the illegally caught fish, that it was easier and less risky to dispose of illegally caught fish at these ports, and that there was generally poor or lax enforcement at these facilities; thus, the risk of being caught was relatively low.[21] The researchers recommended that increasing targeted inspection at these ports, especially if they were large and busy ports, was among the strategies that could be used to stop illegal fishing.

A study examining the patterns of illegal recreational fishing in the no-take zones in the Great Barrier Reef Marine Park in Australia also used the risky facilities framework to explain the spatial distribution of this illegal activity in marine-protected areas. The researchers found that 80 percent of all the illegal recreational fishing happened in 20 percent of no-take zones, where half of all illegal activities took place in only 10 such zones accounting for 6.2 percent of all no-take zones in the area. The researchers also found that these high concentrations were related to the target availability and lack of guardianship. These areas also required less effort to access and "exhibited high levels of local participation"—thus, lending overall support to the assumption that opportunity plays a role when deciding where to carry out illegal recreational fishing.[22]

What Other Strategies Are Available to Analyze Illegal Fishing?

Geographic profiling is an analytical strategy that is used by crime analysts to not only identify crime locations but also use these locations to identify the offenders. This strategy is based on the assumption that information about crime locations can be used to extract information about the victims, offenders, and their interactions.[23] This analytical strategy is based on empirically validated assumptions that offenders will not travel far to commit crime and that they are likely to carry out their criminal activities in areas they are most familiar with and where they can gain easy access to potential victims and suitable exit points.

The *designing against crime* strategy focuses on the design and manufacturing features of a product and how these designs (such as packaging) can be utilized to make the product "objectively harder, riskier or less rewarding for the offender to exploit, or making them perceived as such by the offender."[24] Eco-labeling products is an example of designing fisheries products against crime, so is using blockchain technologies to track the fish from "bait to plate."

The *market reduction approach* argues that disrupting stolen goods markets would be another way of stopping the theft of these goods. This strategy relies on the systematic analysis of theft data, combined with the analysis of interview data collected from offenders and agencies, to understand the buying and selling patterns of demanded goods. Interventions devised to stop the sale of the illegally obtained goods would then focus on the reduction of the demand for these goods by either (a) changing the nature of the markets or (b) changing the nature of the products that are most frequently stolen (through redesigning or other strategies).[25]

Measuring Displacement and Diffusion of Crime Control Benefits

One of the major criticisms of implementing targeted interventions is its potential to *displace* the crime from one location, target, offender, and method to another. Environmental criminologists have identified five types of displacement, which include the following:

1. Geographical displacement (when crime moves from one location to another)

2. Temporal displacement (when crime moves from one time to another)
3. Target displacement (when crime moves from one target to another)
4. Tactical displacement (methods of crime commission change)
5. Crime type displacement (one type of crime is substituted for another)

While many evaluations of targeted intervention in the field of criminal justice have identified insignificant displacement of crime,[26] it is, nevertheless, important that the analyses that examine illegal fishing take into account the possibility of displacement and make efforts to measure it. For example, the weighted displacement quotient analytical technique can be used to measure the geographic displacement of illegal fishing.[27] The same analytical technique can be applied to measure *diffusion of crime control benefits* (i.e., determining whether the prevention strategy implemented in the target area has had any spillover effect). For example, if the targeted intervention involves increasing MCS in the exclusive economic zone of a country or at a port, it may result in diffusion of benefits at these locations: the MCS, if targeted on the illegal fishing of CRAVED species, may result in the reduction of illegal fishing of other commercial species. Targeted MCS will also likely spill into reducing other maritime crimes. For example, instances of human trafficking and migrant smuggling by using fishing vessels have been recorded by the UN Office on Drugs and Crime, which found that while fishers may not be directly engaged in organized trafficking or smuggling activities, they are known to accept bribes to smuggle migrants from developing countries into Europe and the United States. The same study also found that the use of fishing vessels in the illicit traffic in drugs (such as cocaine, heroin, cannabis, and amphetamine-type stimulants) is also widespread. During a three-year period, from 2007 to 2010, maritime patrol has seized a total of 52.3 tonnes of cocaine from vessels, of which 44.5 percent were seized from fishing vessels.[28] Some illegal fishing vessels have also been implicated in fuel smuggling and trafficking of fishing crew,[29] as well as firearms trafficking, maritime piracy, and terrorism.[30] Therefore, targeted interventions will likely lead to diffusion of benefits.

WHAT MORE CAN BE DONE?

In Chapter 4, we provided a general overview on situational crime prevention, and the 25 techniques that fall under five headings, with examples of how these have been used to reduce crime in general. The chapter also provided general guidelines on how these same techniques can be used to deal with illegal fishing. Chapter 6 provided specific examples of promising prevention strategies that have been implemented to stop illegal fishing, many of which are situational in nature, albeit not explicitly labeled as such. Later we provide additional discussion about what can be done to deal with illegal fishing.

Improve Fisheries Management and Formal Surveillance Capacity through Global Partnerships

Countries that have a significant illegal fishing problem should work closely with those that have the "best" MCS measures in place (as shown in analyses in Chapter 5) in an effort to learn from their best practices, forge partnerships, and design evidence-based strategies.

Norway, for example, is one of the "best" countries in terms of its performance on MCS. It was the "best" country when its "observer schemes" and "catch inspection schemes" were evaluated. Norway's catch inspection programs incorporate physically inspecting catches at ports, and the coast guard conducting catch inspections at sea. Further, these inspections aim not only to identify illegally caught species but also to ensure that the correct gear was used, the vessels were on the permitted fishing grounds, and the fish were not dumped.[31]

Namibia received the "best" score for "controlling access to stop illegal fishing" because the country is not only able to control the landing of illegally caught fish at its ports but it also does not allow its ports to be used for transshipments. Moreover, Namibia is one of the very few countries in the world, along with Australia, Canada, and Chile, that has significant penalties for illegal fishing, thus discouraging such activity within its waters.[32]

Using the "TrackWell" VMS is part of the reason why Iceland earned the "best" score on "vessel monitoring schemes." "TrackWell" is an application used by the country's monitoring centers to watch the fishing operations at sea. The program provides "real-time graphical updates of vessel locations on a computerized map, handles catch and activity reports and gives easy access to information on each vessel."[33]

Enhance International Cooperation to Deal with the Problem of POCs

The role of POCs in facilitating illegal fishing should not be underestimated. In this book, we showed that the 10 POCs accounted for 37 percent of all the port visits by detectable illegal fishing vessels. Therefore, "screening the exits"—that is, the ports of the countries that have significant fishing within their waters—should be employed more rigorously to achieve success in reducing illegal fishing. As such, the Ports States Measures Agreement is one of the best tools available to deal with this problem.

Importantly, the international community should raise awareness and encourage not only the major fishing but also other coastal countries (especially those that have POCs, such as Singapore) to strengthen their port controls. This is especially important in cases when these countries are not part of any international fisheries management agreements. In an effort to gain their cooperation, these countries can be provided resources and incentives so that they can improve the control measures in their ports, especially those used by illegal fishing vessels to offload their catch.

Focus on Safeguarding the Species That Are Highly Commercial Internationally

While illegal fishing vessels may target many species, only a few of these species are highly sought after, as demonstrated by the analyses in Chapter 5. Strengthening regulations and international trade controls on these "hot species," as well as devising well-planned and focused enforcement measures, will prove vital in achieving overall reductions in illegal fishing. In addition to monitoring the threatened and most vulnerable fish species (such as bluefin tuna), each country's fisheries authority should prioritize the controls over the species that are abundant and those that are highly commercial internationally that are found within its EEZ, because illegal fishers are much more likely to target these species. While countries may focus their monitoring and surveillance efforts on the few, less abundant and more threatened species, illegal fishers take advantage of these circumstances to target the species that are far more abundant, accessible, and easily "disposable," therefore making fortunes fishing illegally—without much fear of being apprehended.

Prioritize Response Strategies Based on One's Geographic Location

In Chapter 5, we provided a basic geographic breakdown to understand the regional patterns of illegal fishing. These analyses showed, for example, that MCS measures varied regionally and had the strongest deterrent effect in North and South Americas, while these measures were relatively weaker in Southeast and East Asia. As such, retaining strong MCS in the former would prove more effective than attempting to make other changes, such as spending more resources on safeguarding the highly commercial species found within their EEZs. Meanwhile, safeguarding the highly commercial species (e.g., through rigorous requirements for access agreements or licensing to catch these species) needs to be a priority for countries in the Southeast and East Asia, as it was the availability of these species within their EEZs that made these countries especially vulnerable to illegal fishing. Additionally, efforts in blocking access to POCs should be much more rigorous in South and Southeast Asia than, for example, in South America, where, in fact, while such ports existed, access to these POCs had the weakest effect on the overall degree of illegal fishing in the region.

CONCLUDING REMARKS

This chapter reemphasized the importance of looking at the problem illegal fishing through a criminological lens. Shaping illegal fishing as a crime problem, and devising strategies by using the tools that derive from these theoretical underpinnings, is a promising direction to take to deal with the problem in the future.

This chapter also provided a broad overview of the analytical strategies, widely used by crime scientists to study everyday crimes, which fisheries managers, in collaboration with crime scientists, can use to examine the crime of illegal fishing. Criminologists have already used some of these strategies to study illegal fishing, and they have been able to demonstrate the relevance and utility of these strategies in the context of this crime. These studies provide a starting point to the discourse of what works, what is likely to work, and what remains to be tested.

Lastly, this chapter provided additional recommendations that derived from the finding of the analyses of the data in Chapter 5 to illustrate that prevention strategies are likely to work if they are based on data-driven analyses. Notably, these analyses emphasized the importance of focusing on microlevel factors when studying illegal fishing and avoiding overgeneralizations of what works to stop this crime, with one important implication of this being that small situational efforts are likely to lead to significant global changes.

Notes

Chapter 1

1. Sahrhage, D. & Lundbeck, J. (1992). *A History of Fishing*. Berlin, Heidelberg: Springer-Verlag.

2. Ibid.

3. Campbell, S. K. & Butler, V. L. (2010). Archaeological evidence for resilience of Pacific Northwest salmon populations and the socioecological system over the last ~7500 years. *Ecology and Society, 15*(1): 17.

4. Sahrhage, D. & Lundbeck, J. (1992). *A History of Fishing*. Berlin, Heidelberg: Springer-Verlag.

5. Watson, R., Hoshino, E., Beblow, J., Revenga, C., Kura, Y. & Kitchingman, A. (2004). Fishing gear associated with global marine catches. *Fisheries Centre Research Reports, 12*(6): 3–32.

6. Pauly, D., Watson, R. & Alder, J. (2005). Global trends in world fisheries: Impacts on marine ecosystems and food security. *Philosophical Transactions of the Royal Society, 360*: 5–12.

7. Radcliffe, W. (1921). *Fishing from the Earliest Times*. London, UK: John Murray.

8. Adovasio, J. M., Soffer, O. & Klima, B. (1996). Upper Palaeolithic fiber technology: Interlaced woven finds from Pavlov I, Czech Republic, c. 26,000 years ago. *Antiquity, 70*(269): 526–534.

9. Pitcher, T. J. & Lam, M. E. (2015). Fish commoditization and the historical origins of catching fish for profit. *Maritime Studies, 14*(2): 1–19.

10. Campbell, S. K. & Butler, V. L. (2010). Archaeological evidence for resilience of Pacific Northwest salmon populations and the socioecological system over the last ~7500 years. *Ecology and Society, 15*(1): 17.

11. Sahrhage, D. & Lundbeck, J. (1992). *A History of Fishing*. Berlin, Heidelberg: Springer-Verlag.

12. Brewer, D. J. & Friedman, R. F. (1990). *Fish and Fishing in Ancient Egypt*. Cairo, Egypt: The American University in Cairo Press.

13. Pitcher, T. J. & Lam, M. E. (2015). Fish commoditization and the historical origins of catching fish for profit. *Maritime Studies, 14*(2): 1–19.

14. Sahrhage, D. & Lundbeck, J. (1992). *A History of Fishing*. Berlin, Heidelberg: Springer-Verlag.

15. Ibid.

16. Longo, S. B. & Clark, B. (2012). The commodification of bluefin tuna: The historical transformation of the Mediterranean fishery. *Journal of Agrarian Change, 12*(2–3): 204–226.

17. Curtis, R. I. (2005). Sources for production and trade of Greek and Roman processed fish. In Bekker-Nielsen, T. (Ed.). *Ancient Fishing and Fish Processing in the Black Sea Region*. Esbjerg, Denmark: Aarhus University Press.

18. Sahrhage, D. & Lundbeck, J. (1992). *A History of Fishing*. Berlin, Heidelberg: Springer-Verlag.

19. Concoran, T. H. (1963). Roman fish sauces. *The Classical Journal, 58*(5): 204–210.

20. Kurlansky, M. (2002). *Salt: A World History*. New York, NY: Walker and Company.

21. Braund, S. M. (Trans.). (2004). *Juvenal and Persius. Loeb Classical Library 91*. Cambridge, MA, USA: Harvard University Press. Cited in Pitcher, T. J. & Lam, M. E. (2015). Fish commoditization and the historical origins of catching fish for profit. *Maritime Studies, 14*(2): 1–19.

22. Sahrhage, D. & Lundbeck, J. (1992). *A History of Fishing*. Berlin, Heidelberg: Springer-Verlag.

23. Ibid.

24. Pitcher, T. J. & Lam, M. E. (2015). Fish commoditization and the historical origins of catching fish for profit. *Maritime Studies, 14*(2): 1–19.

25. Sahrhage, D. & Lundbeck, J. (1992). *A History of Fishing*. Berlin, Heidelberg: Springer-Verlag.

26. This league was formed in the thirteenth century under the leadership of the trading city-state of Lübeck to defend the interests of the merchant class in the northern German towns in the Baltic region.

27. Kurlansky, M. (1997). *Cod: A Biography of the Fish That Changed the World*. New York, NY: Penguin Books.

28. Sahrhage, D. & Lundbeck, J. (1992). *A History of Fishing*. Berlin, Heidelberg: Springer-Verlag.

29. Pitcher, T. J. & Lam, M. E. (2015). Fish commoditization and the historical origins of catching fish for profit. *Maritime Studies, 14*(2): 1–19.

30. Cushing, D. H. (1988). *The Provident Ocean*. London, UK: Cambridge University Press.

31. Kurlansky, M. (1997). *Cod: A Biography of the Fish That Changed the World*. New York, NY: Penguin Books.

32. Sahrhage, D. & Lundbeck, J. (1992). *A History of Fishing*. Berlin, Heidelberg: Springer-Verlag.

33. Kurlansky, M. (1997). *Cod: A Biography of the Fish That Changed the World*. New York, NY: Penguin Books, p. 24.

34. Hoffman, R. C. (1996). Economic development and aquatic ecosystems in medieval Europe. *The American Historical Review, 101*(3): 631–669.

35. Sahrhage, D. & Lundbeck, J. (1992). *A History of Fishing*. Berlin, Heidelberg: Springer-Verlag.

36. Ibid.

37. Kurlansky, M. (1997). *Cod: A Biography of the Fish That Changed the World.* New York, NY: Penguin Books.

38. Roberts, C. (2007). *The Unnatural History of the Sea.* Washington, DC, USA: Island Press.

39. Pitcher, T. J. & Lam, M. E. (2015). Fish commoditization and the historical origins of catching fish for profit. *Maritime Studies, 14*(2): 1–19.

40. Kurlansky, M. (1997). *Cod: A Biography of the Fish That Changed the World.* New York, NY: Penguin Books.

41. Hilborn, R. & Hilborn, U. (2012). *Overfishing: What Everyone Needs to Know.* New York, NY: Oxford University Press.

42. Ibid.

43. Kurlansky, M. (1997). *Cod: A Biography of the Fish That Changed the World.* New York, NY: Penguin Books.

44. Bockstoce, J. (1986). *Whales, Ice, and Men: The History of Whaling in the Western Arctic.* Seattle, WA: University of Washington Press.

45. Sahrhage, D. & Lundbeck, J. (1992). *A History of Fishing*. Berlin, Heidelberg: Springer-Verlag.

46. Duplés-Agier, H. (1852). Ordonnances inédites de Philippe le Bel et de Philippe le Long Sur La Police de la Pêche Fluviale. *Bibliothèque de l'Ecole des Chartes, 14*: 49. Cited in Hoffman, R. C. (1996). Economic development and aquatic ecosystems in medieval Europe. *The American Historical Review, 101*(3): 631–669.

47. Brandt, A. V. (1972). *Fish Catching Methods of the World. Revised and Enlarged.* London, UK: Fishing News (Books) Ltd.

48. Cushing, D. H. (1988). *The Provident Ocean.* London, UK: Cambridge University Press.

49. Sahrhage, D. & Lundbeck, J. (1992). *A History of Fishing*. Berlin, Heidelberg: Springer-Verlag.

50. Pitcher, T. J. & Lam, M. E. (2015). Fish commoditization and the historical origins of catching fish for profit. *Maritime Studies, 14*(2): 1–19.

51. Wimpenny, R. S. (1953). *The Plaice: Being the Buckland Lectures for 1949.* London, UK: Edward Arnold Publishers.

52. Cushing, D. H. (1988). *The Provident Ocean.* London, UK: Cambridge University Press.

53. Longo, S. B. & Clark, B. (2012). The commodification of bluefin tuna: The historical transformation of the Mediterranean fishery. *Journal of Agrarian Change, 12*(2–3): 204–226.

54. Pitcher, T. J. & Lam, M. E. (2015). Fish commoditization and the historical origins of catching fish for profit. *Maritime Studies, 14*(2): 1–19.

55. Ibid.

56. Ibid.

57. Alward, G. L. (1932). *The Sea Fisheries of Great Britain and Ireland: A Record of the Development of the Fishing Industry and Its Worldwide Ramifications. Notes from*

the Author's Diary, 1854–1928, with Charts, Photographs and Illustrations. Grimsby, UK: Albert Gait. Cited in Pitcher, T. J. & Lam, M. E. (2015). Fish commoditization and the historical origins of catching fish for profit. *Maritime Studies, 14*(2): 1–19.

58. Pitcher, T. J. & Lam, M. E. (2015). Fish commoditization and the historical origins of catching fish for profit. *Maritime Studies, 14*(2): 1–19.

59. Takayam, S. (1963). Mechanization of fishing fleets in Japan. In Nui, L. B. & Jan-Old Trrmg, J.-O. (Eds.). *Indo-Pacific Fisheries Council Proceedings.* 10th Session, Seoul, Republic of Korea, Section II Symposium Papers. Bangkok: IPFC Secretariat, FAO Regional Office for Asia and the Far East, pp. 1–117.

60. Morgan, G. R. & Staples, D. J. (2006). *The History of Industrial Marine Fisheries in Southeast Asia.* Bangkok, Thailand: Food and Agriculture Organization of the United Nations Regional Office for Asia and the Pacific.

61. Jia, J. & Chen, J. (2001). *Sea Farming and Sea Ranching in China.* FAO Fisheries Technical Paper 418. Rome: FAO.

62. Sahrhage, D. & Lundbeck, J. (1992). *A History of Fishing.* Berlin, Heidelberg: Springer-Verlag.

63. Ibid.

64. Gelchu, A. & Pauly, D. (2007). Growth and distribution of port-based global fishing effort within countries' EEZs from 1970 to 1995. *Fisheries Centre Research Reports, 15*(4): 1–86.

65. Gulland, J. A. (1974). *The Management of Marine Fisheries.* Bristol: Scientechnica (Publishers) Ltd.

66. Doughty, R. W. & Carmichael, V. (2011). *The Albatross and the Fish: Linked Lives in the Open Seas.*

67. Thiele, W. (1999). *Global Trends in Fishing Technology and Their Effects on Fishing Power and Capacity.* Busan, Republic of Korea: FAO Regional Workshop on the Effects of Globalization and Deregulation of Marine Capture Fisheries in Asia.

68. Chidambaram, K. (1963). Development of mechanized fishing fleets in the Indo-Pacific council region. In Nui, L. B. & Jan-Old Trrmg, J.-O. (Eds.). *Indo-Pacific Fisheries Council Proceedings.* Section II, Symposium Papers. Bangkok: IPFC Secretariat, FAO Regional Office for Asia and the Far East, pp. 2–27.

69. Gelchu, A. & Pauly, D. (2007). Growth and distribution of port-based global fishing effort within countries' EEZs from 1970 to 1995. *Fisheries Centre Research Reports, 15*(4): 1–86.

70. Solecki, J. J. (1979). A review of USSR fishing industry. *Ocean Management, 5*: 97–123.

71. Anon. (2002). *Assessment Report on Fisheries and Fisheries Related Species and Habitat Issues.* Bergen, Norway: North Sea Conference. Cited in Gelchu, A. & Pauly, D. (2007). Growth and distribution of port-based global fishing effort within countries' EEZs from 1970 to 1995. *Fisheries Centre Research Reports, 15*(4): 1–86.

72. APO (Asian Productivity Organization). (1988). *Fishing industry in Asia and Pacific.* Hong Kong, China: Medical International Limited; Asada, Y., Hirasawa, Y. & Nagasaki, F. (1983). *Fisheries Management in Japan.* FAO Fisheries Technical Paper 238. Rome: FAO.

73. Jia, J. & Chen, J. (2001). *Sea Farming and Sea Ranching in China.* FAO Fisheries Technical Paper 418. Rome: FAO.

74. Gelchu, A. & Pauly, D. (2007). Growth and distribution of port-based global fishing effort within countries' EEZs from 1970 to 1995. *Fisheries Centre Research Reports, 15*(4): 1–86.

75. Brian, R. (1985). Recent developments in Australian fishing industry. In Cleveland, B. C. (Ed.). *Adjustments to Changes in Fisheries Law and Economics.* FAO Fisheries Technical Paper 269. Suppl. Rome: FAO, pp. 21–26.

76. Johnson, V. (1992). Artisanal fishermen and intermediaries in Mozambique. In Tvedten, I. & Hersoug, B. (Eds.). *Fishing for Development: Small-Scale Fisheries in Africa.* Sweden: Nordiska Afrikainstitutet (The Scandinavian Institute of African Studies), pp. 70–90.

77. Njofonjou, O. & Njock, J. C. (2000). African fisheries: Major trends and diagnostic of the 20th century. In *Microbehavior and Macroresults: Proceedings of the Tenth Biennial Conference of the International Institute for Fisheries Economics and Trade,* July 10–14. Corvallis, OR: Oregon State University, Compiled by Richard S. Johnston & Ann L. Shriver. International Institute of Fisheries Economics and Trade (IIFET), Corvallis, 2001.

78. Tvedten, I. & Hersoug, B. (Eds.). (1992). *Fishing for Development: Small-Scale Fisheries in Africa.* Sweden: Nordiska Afrikainstitutet (The Scandinavian Institute of African studies).

79. Lawson, R. M. & Kwei, E. R. (1974). *African Entrepreneurship and Economic Growth: A Case Study of the Fishing Industry of Ghana.* Accra: Ghana Universities Press.

80. Gelchu, A. & Pauly, D. (2007). Growth and distribution of port-based global fishing effort within countries' EEZs from 1970 to 1995. *Fisheries Centre Research Reports, 15*(4): 1–86.

81. Gulland, J. A. (1974). *The Management of Marine Fisheries.* Bristol: Scientechnica (Publishers) Ltd.

82. Pitcher, T. J. & Lam, M. E. (2015). Fish commoditization and the historical origins of catching fish for profit. *Maritime Studies, 14*(2): 1–19.

83. Pauly, D., Christensen, V., Guénette, S., Pitcher, T. J., Sumaila, U. R., Walters, C. J., Watson, R. & Zeller, D. (2002). Toward sustainability in world fisheries. *Nature, 418*(8): 689–695.

84. MacSweet, I. M. (1983). *Markets for Fish and Fishery Products in Eastern Europe.* FAO Fisheries Technical Paper 241 FIIU/T241. Rome: FAO; Garcia, S. M. & Newton, C. (1997). Current situation, trends and prospects in world capture fisheries. In Pikitch, E. L., Huppert, D. D. & Sissenwine, M. P. (Eds.). *Global Trends: Fisheries Management.* Seattle, WA: American Fisheries Society Symposium 20, 14–16 June 1994, pp. 3–27.

85. Report on the Status of Groundfish Stocks in the Canadian Northwest Atlantic. Atlantic Stock Assessment Secretariat, Department of Fisheries and Oceans, June 1994, p. 19.

86. Oceans, Ltd. (1990). *Reproductive Success in Atlantic Cod (Gadus morhua): The Potential Impact of Trawling.* A Report Prepared for the Newfoundland Inshore Fisheries Association.

87. Kurlansky, M. (1997). *Cod: A Biography of the Fish That Changed the World.* New York, NY: Penguin Books.

88. Report on the Status of Groundfish Stocks in the Canadian Northwest Atlantic. Atlantic Stock Assessment Secretariat, Department of Fisheries and Oceans, June 1994, p. 19.

89. Ibid.

90. Hilborn, R. & Hilborn, U. (2012). *Overfishing: What Everyone Needs to Know.* New York, NY: Oxford University Press.

91. Bockstoce, J. (1986). *Whales, Ice, and Men: The History of Whaling in the Western Arctic.* Seattle, WA: University of Washington Press.

92. Jackson, J. B. C., et al. (2001). Historical overfishing and the recent collapse of coastal ecosystems. *Science, 293*(27): 629–636.

93. Marine Conservation Society. (n.d.). *How Fish Are Caught.* www.mcsuk.org/downloads/fisheries/Fishing_Methods.pdf.

94. Ibid.

95. Sebater, M. (2007). *Prehistoric, Historic and Current Reef Fish Utilization and the Consequence of Human Population Growth on Fish Density, Biomass and Community Composition in American Samoa.* Bio. Rep. Series No. 2007-01. American Samoa: Department of Marine and Wildlife Resources.

96. FAO (Food and Agriculture Organization of the United Nations). (n.d.). Fisheries section: Glossary. *Fishing Industry.* www.fao.org/fi/glossary/default.asp.

97. Dalzell, P., Adams, T. & Polunin, N. (1997). Coastal fisheries in the Pacific Islands. *Oceanographic Literature Review, 44* (5): 516.

98. Williamson, M. J. (2013). *New Report Highlights Recreational Fishing's Broad Economic and Conservation Impact.* American Sportfishing Association Press Release. Accessed October 7, 2013, http://www.southwickassociates.com/blog/2013/01/10/new-report-highlights-recreational-fishings-broad-economic-and-conservation-impact/.

99. FAO (Food and Agriculture Organization of the United Nations). (2014). *FAO Yearbook.* United Nations: Fisheries and Aquaculture Statistics.

100. Ibid.

101. Ibid.

102. Ibid.

103. Ibid.

104. Ibid.

105. Ibid.

106. Ibid.

107. Ibid.

108. Ibid.

109. FAO (Food and Agriculture Organization of the United Nations). (2016). *The State of World Fisheries and Aquaculture. Contributing to Food Security and Nutrition for All.* Rome.

110. UNODC (United Nations Office on Drugs and Crime). (2011). *Transnational Organized Crime in the Fishing Industry: Focus on Trafficking in Persons, Smuggling of Migrants and Illicit Drugs Trafficking.* Vienna: United Nations.

111. FAO (Food and Agriculture Organization of the United Nations). (2016). *The State of World Fisheries and Aquaculture. Contributing to Food Security and Nutrition for All.* Rome.

112. Love, P. (2010). *Fisheries: While Stocks Last.* Organization for Economic Cooperation and Development.

113. Watson, R., Hoshino, E., Beblow, J., Revenga, C., Kura, Y. & Kitchingman, A. (2004). Fishing gear associated with global marine catches. *Fisheries Centre Research Reports, 12*(6): 3–32.

114. Marine Conservation Society. (n.d.). *How Fish Are Caught*. www.mcsuk.org/downloads/fisheries/Fishing_Methods.pdf.

115. FAO (Food and Agriculture Organization of the United Nations). (n.d.). Fisheries section: Glossary. *Fishing Industry*. www.fao.org/fi/glossary/default.asp.

116. Ibid.

117. Fishing Vessel types. Trawlers. Technology Fact Sheets. In: *FAO Fisheries and Aquaculture Department* [online]. Rome. Updated 9 May 2001. [Cited 15 October 2013]. http://www.fao.org/fishery/vesseltype/10/en.

118. Ibid.

119. FAO (Food and Agriculture Organization of the United Nations). (2016). *FAO Fisheries and Aquaculture Report No. 932*. United Nations. FIRF/R932 (En).

120. Ibid.

121. UNODC (United Nations Office on Drugs and Crime). (2011). *Transnational Organized Crime in the Fishing Industry: Focus on Trafficking in Persons, Smuggling of Migrants and Illicit Drugs Trafficking*. Vienna: United Nations.

122. Monterey Bay Aquarium. (2013). *Fishing and Farming Methods*. Accessed October 15, 2013, https://www.seafoodwatch.org/ocean-issues/fishing-and-farming-methods.

123. Fishing Vessel types. Trawlers. Technology Fact Sheets. In: *FAO Fisheries and Aquaculture Department* [online]. Rome. Updated 9 May 2001. [Cited 15 October 2013]. http://www.fao.org/fishery/vesseltype/10/en.

124. Monterey Bay Aquarium. (2013). *Fishing and Farming Methods*. Accessed October 15, 2013, https://www.seafoodwatch.org/ocean-issues/fishing-and-farming-methods.

125. Ibid.

126. Ibid.

127. Ibid.

128. Fishing Vessel types. Trawlers. Technology Fact Sheets. In: *FAO Fisheries and Aquaculture Department* [online]. Rome. Updated 9 May 2001. [Cited 15 October 2013]. http://www.fao.org/fishery/vesseltype/10/en.

129. Monterey Bay Aquarium. (2013). *Fishing Vessel Fact Cards*. Accessed October 15, 2013, http://www.montereybayaquarium.org/cr/cr_seafoodwatch/content/media/MBA_SeafoodWatch_GearFactCards.pdf.

130. Clark, M. R. & Koslow, J. A. (2007). Impacts of fisheries on seamounts. In Pitcher, T. J., Morato, T., Hart, P. J. B., Clark, M. R., Haggan, N., & Santos, R. S. (Eds.). *Seamounts: Ecology, Fisheries, and Conservation*. Fish and Aquatic Resources Series, 12. Oxford, UK: Blackwell Publishing.

131. Pitcher, T. J. & Lam, M. E. (2015). Fish commoditization and the historical origins of catching fish for profit. *Maritime Studies, 14*(2): 1–19.

Chapter 2

1. The term "IUU fishing" was first defined in 2001 in the UN Food and Agriculture Organization (FAO)'s International Plan of Action to Prevent, Deter and Eliminate Illegal, Unreported and Unregulated Fishing (IPOA-IUU).

2. These refer to the waters of the states that are archipelagos, such as Japan, the Philippines, and Indonesia.

3. FAO Code of Conduct. (1995). *Adopted at the 28th Session of the FAO Conference.* Rome, Italy, October 31.

4. Bailey, C. (1997). Lessons from Indonesia's 1980 trawler ban. *Marine Policy, 21* (3): 225–235.

5. EJF (Environmental Justice Foundation). (2007). *Illegal Driftnetting in the Mediterranean.* London, UK: Environmental Justice Foundation.

6. Phelps Bondaroff, T. N., Reitano, T. & van der Werf, W. (2015). *The Illegal Fishing and Organized Crime Nexus: Illegal Fishing as Transnational Organized Crime.* The Global Initiative against Transnational Crime and the Black Fish. https://globalinitiative.net/wp-content/uploads/2015/04/the-illegal-fishing-and-organised-crime-nexus-1.pdf.

7. House of Ocean. (2015). IUU: Is It a Bird or a Plane? Is It Illegal Fishing, Unregulated Fishing or Crime? Look to the Fish Stocks Agreement for Answers. Accessed October 6, 2017, https://houseofocean.org/2015/10/03/iuu-is-it-a-bird-or-a-plane-is-it-illegal-fishing-unregulated-fishing-or-crime-look-to-the-fish-stocks-agreement-for-answers/.

8. Ibid.

9. Ibid.

10. NEAFC (North East Atlantic Fisheries Commission). (2009). *Scheme of Control and Enforcement.* Article 29.

11. Palma, M. A., Tsamenyi, M. & Edeson, W. (2010). *Promoting Sustainable Fisheries: The International Legal and Policy Framework to Combat Illegal, Unreported and Unregulated Fishing.* Leiden, Netherlands: Martinus Nijhoff Publishers.

12. EJF (Environmental Justice Foundation). (2007). *Illegal Driftnetting in the Mediterranean.* London, UK: Environmental Justice Foundation.

13. EJF (Environmental Justice Foundation). (2005). *Pirates and Profiteers: How Pirate Fishing Fleets Are Robbing People and Oceans.* London, UK: Environmental Justice Foundation.

14. Falaye, A. E. (2008). Illegal Unreported Unregulated (IUU) Fishing in West Africa (Nigeria & Ghana). *Marine Resources Assessment Group Ltd, London.*

15. HSTF (High Seas Task Force). (2006). *Closing the Net: Stopping Illegal Fishing on the High Seas.* Final report of the Ministerially-led Task Force on IUU Fishing on the High Seas. London, UK. www.high-seas.org.

16. MRAG. (2008). *Study and Analysis of the Status of IUU Fishing in the SADC Region and an Estimate of the Economic, Social and Biological Impacts.* Volume 2-Main Report. https://mrag.co.uk/services/illegal-unreported-and-unregulated-fishing?page=1.

17. Pew Charitable Trusts. (December 10, 2013). *How to End Illegal Fishing.* https://www.pewtrusts.org/en/research-and-analysis/reports/2013/12/10/how-to-end-illegal-fishing.

18. HSTF (High Seas Task Force). (2006). *Closing the Net: Stopping Illegal Fishing on the High Seas.* Final report of the Ministerially-led Task Force on IUU Fishing on the High Seas. London, UK. www.high-seas.org.

19. EJF (Environmental Justice Foundation). (2005). *Pirates and Profiteers: How Pirate Fishing Fleets Are Robbing People and Oceans.* London, UK: Environmental Justice Foundation.

20. Sumaila, U. R., Alder, J. & Keith, H. (2006). Global scope and economics of illegal fishing. *Marine Policy, 30*(6): 696–703.

21. Pitcher, T., Kalikoski, D. & Pramod, G. (Eds.). (2006). Evaluations of compliance with the FAO (UN) Code of Conduct for Responsible Fisheries. *Fisheries Centre Research Reports, 14*(2): 1191.

22. Agnew, D. J., Pearce, J., Pramod, G., Peatman, T., Watson, R., Beddington, J. R. & Pitcher, T. J. (2009). Estimating the worldwide extent of illegal fishing. *PLoS ONE, 4* (2): e4570. DOI: 10.1371/journal.pone.0004570.

23. Commission for the Conservation of Antarctic Marine Living Resources (CCAMLR). (2004). *Report of the Twenty-Third Meeting of the Commission.* Hobart, Australia, October 25–November 4. CCAMLR-XXIII, Tasmania, Australia, at 31.

24. Hoydal, K. (2004). IUU fishing in NEAFC: How big is the problem and what have we done? *Fish Piracy: Combating Illegal, Unreported and Unregulated Fishing.* Paris: OECD, p. 161.

25. Yumiko, K., Revenga, C., Hoshino, E. & Mock, G. (2004). *Fishing for Answers: Making Sense of the Global Fish Crisis.* Washington, DC: World Resources Institute. http://www.wri.org/sites/default/files/pdf/fishanswer_fulltext.pdf.

26. Freitas, B. (2015). *Illegal Fishing: Which Fish Species Are at Highest Risk from Illegal and Unreported Fishing?* Washington, DC: World Wildlife Fund.

27. Ibid.

28. Petrossian, G. & Clarke, R. V. (2014). Explaining and controlling illegal commercial fishing: An application of the CRAVED theft model. *British Journal of Criminology, 54*(1): 73–90.

29. Petrossian, G., Weis, J. & Pires, S. F. (2015). Factors affecting crab and lobster species subject to IUU fishing. *Ocean and Coastal Management, 106*: 29–34. DOI: 10.1016/j.ocecoaman.2015.014.

30. United Nations. (2008). *Conference of Parties to the United Nations Convention against Transnational Organized Crime.* Report of the Conference of Parties to the United Nations Convention against Transnational Organized Crime on Its Fourth Session, Vienna, October 8–17. CTOC/COP/2008/19, December 1, para. 210.

31. Chen, T. (2012). Easy pickings for abalone smugglers. *Wall Street Journal,* China Real Time. http://blogs.wsj.com/chinarealtime/2012/03/30/easy-pickings-for-abalone-smugglers/.

32. Shaw, M. (2014). Transnational organized crime in Africa. In Albanese, J. S. & Reichel, P. L. (Eds.). *Transnational Organized Crime: An Overview from Six Continents.* Los Angeles, CA: Sage Publications, pp. 93–115.

33. Warchol, G. & Harrington, M. (2016). Exploring the dynamics of South Africa's illegal abalone trade via routine activities theory. *Trends in Organized Crime, 19*: 21–41.

34. Poling, G. B. & Cronin, C. (2017). *Illegal, Unreported and Unregulated Fishing as a National Security Threat.* Center for Strategic and International Studies in cooperation with National Geographic Pristine Seas. https://www.csis.org/analysis/illegal-unreported-and-unregulated-fishing-national-security-threat.

35. Putt, J. & Nelson, D. (n.d.). Crime in the Australian fishing industry. *Trends and Issues in Crime and Criminal Justice, 366.* https://aic.gov.au/publications/tandi/tandi366.

36. Wyler, L. S. & Sheikh, P. A. (2013). *International Illegal Trade in Wildlife: Threats and US Policy.* Washington, DC: Congressional Research Service Report to Congress.

37. Zabyelina, Y. (2014). The "Fishy Business": A qualitative analysis of the illicit market in black caviar. *Trends in Organized Crime, 17*(3): 181–198.

38. Engler, M. & Parry-Jones, R. (2007). *Opportunity or Threat: The Role of the European Union in Global Wildlife Trade.* Brussels: TRAFFIC Europe.

39. Volkov, V. (2002). *Violent Entrepreneurs: The Use of Force in the Making of Russian Capitalism*. Ithaca, NY and London, UK: Cornell University Press.

40. Van Uhm, D. P. (2016). *Uncovering the Illegal Wildlife Trade: Inside the World of Poachers, Smugglers and Traders*. New York, NY: Springer.

41. Ibid.

42. Telesetsky, A. (2015). Laundering fish in the global undercurrents: Illegal, unreported, and unregulated fishing and transnational organized crime. *Ecology Law Quarterly, 41*(4): 939–996.

43. Österblom, H., Constable, A. & Fukumi, S. (2011). Illegal fishing and the organized crime analogy. *Trends in Ecology and Evolution, 26*(6): 261–262.

44. Telesetsky, A. (2015). Laundering fish in the global undercurrents: Illegal, unreported, and unregulated fishing and transnational organized crime. *Ecology Law Quarterly, 41*(4): 939–996.

45. Africa Progress Panel. (2014). *Launch Remarks—Kofi Annan*. Geneva: Africa Progress Panel Report. https://www.afdb.org/fileadmin/uploads/afdb/Documents/Project-and-Operations/Africa_Progress_Report_2014.PDF.

46. Doughty, R. W. & Carmichael, V. (2011). *The Albatross and the Fish: Linked Lives in the Open Seas*. Austin, TX: University of Texas Press.

47. FAO (Food and Agriculture Organization of the United Nations). (2016). *The State of World Fisheries and Aquaculture*. Rome: FAO.

48. Kaczynski, V. M. & Fluharty, D. L. (2002). European policies in West Africa: Who benefits from fisheries agreements? *Marine Policy, 26*: 75–93.

49. Ibid.

50. Ibid.

51. Standing, A. (2008). *Corruption and Industrial Fishing in Africa*. Bergen, Norway. https://www.cmi.no/publications/file/3188-corruption-andindustrial-fishing-in-africa.pdf.

52. Stiles, M. L., Kagan, A., Shaftel, E. & Lowell, B. (2013). *Stolen Seafood: The Impact of Pirate Fishing on Our Oceans*. Oceana. http://oceana.org/sites/default/files/reports/Oceana_StolenSeafood.pdf.

53. MRAG. (2005). *Review of Impacts of Illegal, Unreported and Unregulated Fishing on Developing Countries*. Final Report. Report prepared for the United Kingdom's Department for International Development with support from the Norwegian Agency for Development Cooperation.

54. EJF (Environmental Justice Foundation). (2005). *Pirates and Profiteers: How Pirate Fishing Fleets Are Robbing People and Oceans*. London, UK: Environmental Justice Foundation.

55. Black, R. (2007). Action urged on illegal fishing. *BBC News*. Accessed October 21, 2013, http://news.bbc.co.uk/2/hi/6602639.stm.

56. Oceana. (2017). *Fishing the Boundaries of Law: How the Exclusivity Clause in EU Fisheries Agreements Was Undermined*. http://usa.oceana.org/publications/reports/fishing-boundaries-law-how-exclusivity-clause-eu-fisheries-agreements-was.

57. Palma, M. A., Tsamenyi, M. & Edeson, W. (2010). *Promoting Sustainable Fisheries: The International Legal and Policy Framework to Combat Illegal, Unreported and Unregulated Fishing*. Leiden, Netherlands: Martinus Nijhoff Publishers.

58. Ibid.

59. Swartz, W., Sumaila, U. R., Watson, R. & Pauly, D. (2010). Sourcing seafood for the three major markets: The EU, Japan and the USA. *Marine Policy, 34*: 1366–1373.

60. European Parliament. (2014). *Illegal, Unreported and Unregulated Fishing: Sanctions in the EU.* Strasbourg: European Parliament. http://www.europarl.europa.eu/RegData/etudes/STUD/2014/529069/IPOL_STU(2014)529069_EN.pdf.

61. Gravitz, M. (2014). *Plundering the Seas: The Damage from Pirate Fishing on US Fishermen & Communities*. Seattle, WA: Marine Conservation Institute. http://www.marine-conservation.org/media/filer_public/filer_public/2014/09/04/mci_iuu_report.pdf.

62. EJF (Environmental Justice Foundation). (2005). *Pirates and Profiteers: How Pirate Fishing Fleets Are Robbing People and Oceans*. London, UK: Environmental Justice Foundation.

63. EJF (Environmental Justice Foundation). (2010). *All at Sea: The Abuse of Human Rights Aboard Illegal Fishing Vessels*. London, UK: Environmental Justice Foundation.

64. UNODC (United Nations Office on Drugs and Crime). (2011). *Transnational Organized Crime in the Fishing Industry: Focus on Trafficking in Persons, Smuggling of Migrants and Illicit Drugs Trafficking*. Vienna: United Nations.

65. Urbina, I. (2015). "Sea slaves": The human misery that feeds pets and livestock. *The New York Times*, July 27.

66. UNODC (United Nations Office on Drugs and Crime). (2011). *Transnational Organized Crime in the Fishing Industry: Focus on Trafficking in Persons, Smuggling of Migrants and Illicit Drugs Trafficking*. Vienna: United Nations.

67. This case study research was cited as "Muhammad/Save the Children Sweden *Commercial Sexual Exploitation of Children: A Situation Analysis of Pakistan* (2005) 49, in UNODC (United Nations Office on Drugs and Crime). (2011). *Transnational Organized Crime in the Fishing Industry: Focus on Trafficking in Persons, Smuggling of Migrants and Illicit Drugs Trafficking*. Vienna: United Nations.

68. ITF (International Transport Workers' Federation). (2006). *Out of Sight, Out of Mind: Seafarers, Fishers & Human Rights*. http://www.itfseafarers.org/files/extranet/-1/2259/HumanRights.pdf.

69. FAO (Food and Agriculture Organization of the United Nations). (2007). *FAO Workshop on Vulnerable Ecosystems and Destructive Fishing in Deep-Sea Fisheries*. FAO Report No. 829. Rome, Italy.

70. Watson, R., Hoshino, E., Beblow, J., Revenga, C., Kura, Y. & Kitchingman, A. (2004). Fishing gear associated with global marine catches. *Fisheries Centre Research Reports, 12*(6): 3–32.

71. Chevallier, R. (2017). Safeguarding Tanzania's coral reefs: The case of illegal blast fishing. *South African Institute of International Affairs, Policy Insights, 46.*

72. Ibid.

73. Cornish, A. S. & McKellar, D. (1998). A history of fishing with explosives and poisons in Hong Kong Waters. *Naga, the ICLARM Quarterly*, July–September, 4–9.

74. Chevallier, R. (2017). Safeguarding Tanzania's coral reefs: The case of illegal blast fishing. *South African Institute of International Affairs, Policy Insights, 46.*

75. Ibid.

76. Caldwell, R. & Fox, H. (2006). Recovery from blast fishing on coral reefs: A tale of two scales. *Ecological Applications, 16*(5): 1631–1635.

77. Ibid.

78. White, A. T., Vogt, H. P. & Arin, T. (2000). The Philippine coral reefs under threat: The economic losses caused by reef destruction. *Marine Pollution Bulletin, 40*: 598–605.

79. Neuwinger, H. D. (2004). Plants used for poison fishing in tropical Africa. *Toxicon, 44*(4): 417–430.

80. UNEP. (1994). *Staff Training Materials for the Management of Marine Protected Areas*. RCU/EAS Technical Report Series No. 4, 216–220.

81. Mak, K. K. W., Yanase, H. & Renneberg, R. (2005). Cyanide fishing and cyanide detection in coral reef fish using chemical tests and biosensors. *Biosensors and Bioelectronics, 20*(12): 2581–2593.

82. Doughty, R. W. & Carmichael, V. (2011). *The Albatross and the Fish: Linked Lives in the Open Seas*. Austin, TX: University of Texas Press.

83. EJF (Environmental Justice Foundation). (2007). *Illegal Driftnetting in the Mediterranean*. London, UK: Environmental Justice Foundation. ISBN No. 1-904523-11-0.

84. Doughty, R. W. & Carmichael, V. (2011). *The Albatross and the Fish: Linked Lives in the Open Seas*. Austin, TX: University of Texas Press.

85. More on this is discussed in the following section.

86. Bryant, D., Burke, L., McManus, J. & Spalding, M. (1998). *Reefs at risk: A map-based indicator of threats to the world's coral reefs*. World Resources Institute, International Center for Living Aquatic Resources Management, World Conservation Monitoring Centre, United Nations Environment Programme, 56 p. https://wriorg.s3.amazonaws.com/s3fs-public/pdf/reefs.pdf?_ga=2.57804330.1209804127.1548434425-2003236149.1548434425.

87. Doughty, R. W. & Carmichael, V. (2011). *The Albatross and the Fish: Linked Lives in the Open Seas*. Austin, TX: University of Texas Press.

88. Ibid.

89. Kuper, J. (2008). *Oceans in Crisis: Situation Status*. https://stopillegalfishing.com/news-articles/oceans-in-crisis-by-jo-kuper-greenpeace-international-5/.

90. Brothers, N. P. (1991). Albatross mortality and associated bait loss in the Japanese longline fishery in the southern ocean. *Biological Conservation, 55*: 255–268.

91. Alverson, D. L., Freeberg, M. K., Murawski, S. A. & Pope, J. G. (1994). *A Global Assessment of Fisheries Bycatch and Discards*. FAO Fisheries Technical Paper No. 339 Rome, FAO.

92. Doughty, R. W. & Carmichael, V. (2011). *The Albatross and the Fish: Linked Lives in the Open Seas*. Austin, TX: University of Texas Press.

93. Gilman, E. (2004). *References on Seabird Bycatch in Longline Fisheries*. Honolulu, HI: Blue Ocean Institute.

94. Paul, L. (1994). *High Seas Driftnetting: The Plunder of the Global Commons*. http://earthtrust.org/archive/dnpaper/contents.html, as cited in EJF (Environmental Justice Foundation). (2007). *Illegal Driftnetting in the Mediterranean*. London, UK: Environmental Justice Foundation. ISBN No. 1-904523-11-0.

95. Doughty, R. W. & Carmichael, V. (2011). *The Albatross and the Fish: Linked Lives in the Open Seas*. Austin, TX: University of Texas Press.

96. Liddick, D. (2014). The dimensions of a transnational crime problem: The case of IUU fishing. *Trends in Organized Crime, 17*: 290–312.

97. Pauly, D., Christensen, F., Dalsgaard, J., Froese, R. & Torres, F. (1998). Fishing down marine food webs. *Science, 279*(5352): 860–863.

98. Ibid.

99. Ferretti, F., Myers, R. A., Serena, F. & Lotze, H. K. (2008). Loss of large predatory sharks from the Mediterranean Sea.*Conservation Biology, 22*(4): 952–964.

100. Pauly, D., Christensen, V., Guénette, S., Pitcher, T. J., Sumaila, U. R., Walters, C. J., Watson, R. & Zeller, D. (2002). Towards sustainability in world fisheries. *Nature, 418*(8): 689–695.

101. Stephens, D. W. & Krebs, J. R. (1986). *Foraging Theory*. Princeton, NJ: Princeton University Press.

102. Neutel, A. M., Heesterbeek, J. A. P. & de Ruiter, P. C. (2002). Stability in real food webs: Weak links in long loops. *Science, 296*: 1120–1123.

103. Jackson, J. B. C., Kirby, M. X., Berger, W. H., Bjorndal, K. A., Botsford, L. M., Bourque, B. J., . . . Hughes, T. P. (2001). Historical overfishing and the recent collapse of coastal ecosystems. *Science, 293*: 629–636.

Chapter 3

1. United Nations. (2005). *Convention on the High Seas*. Done at Geneva on April 29, 1958. Entered into force on September 30, 1962. *Treaty Series,* Vol. 450.

2. Ibid.

3. Osieke, E. (1979). Flags of convenience vessels: Recent developments. *The American Journal of International Law, 73*: 604–627.

4. Llácer, F. J. M. (2003). Open registers: Past, present and future. *Marine Policy, 27* (6): 513–523.

5. Shaughnessy, T. & Tobin, E. (2006). Flags of inconvenience: Freedom and insecurity on the high seas. *Journal of International Law, 5*(1): 1–31.

6. McLeod, W. J. (1964). The flags-of-convenience problem. *South Carolina Law Review, 16*: 409–418.

7. Shaughnessy, T. & Tobin, E. (2006). Flags of inconvenience: Freedom and insecurity on the high seas. *Journal of International Law, 5*(1): 1–31.

8. EJF (Environmental Justice Foundation). (2009). *Lowering the Flag—Ending the Use of Flags of Convenience by Pirate Fishing Vessels*. London, UK: Environmental Justice Foundation.

9. Behnam, A. & Faust, P. (2003). Twilight of flag state control. *Ocean Yearbook Online, 17*(1). DOI: https://doi.org/10.1163/221160003X00096.

10. Ibid.

11. Ibid.

12. Ibid.

13. EJF (Environmental Justice Foundation). (2009). *Lowering the Flag—Ending the Use of Flags of Convenience by Pirate Fishing Vessels*. London, UK: Environmental Justice Foundation.

14. Ibid.

15. Clover, C. (2004). *The End of the Line*. London, UK: Ebury Press, Random House, p. 129.

16. Martini, M. (2013). *Illegal, Unreported and Unregulated Fishing and Corruption*. U4 Anti-Corruption Resource Centre. www.U4.no.

17. Gianni, M. & Simpson, W. (2005). *The Changing Nature of High Seas Fishing: How Flags of Convenience Provide Cover for Illegal, Unreported and Unregulated Fishing*. Australian Government, Department of Agriculture, Fisheries and Forestry, International Transport Workers' Federation, and WWF International.

18. Swan, J. (2002). *Fishing Vessels Operating under Open Registers and the Exercise of Flag State Responsibilities: Information and Options*. FAO Fisheries Circular No. 980, Rome.

19. Agnew, D. J. & Barnes, C. T. (2004). Economic aspects and drivers of IUU fishing: Building a framework in OECD. In *Fish Piracy. Combating Illegal, Unreported and Unregulated Fishing*. Paris: OECD.

20. Gianni, M. & Simpson, W. (2005). *The Changing Nature of High Seas Fishing: How Flags of Convenience Provide Cover for Illegal, Unreported and Unregulated Fishing*. Australian Government, Department of Agriculture, Fisheries and Forestry, International Transport Workers' Federation, and WWF International.

21. Doughty, R. W. & Carmichael, V. (2011). *The Albatross and the Fish: Linked Lives in the Open Seas*. Austin, TX: University of Texas Press.

22. George, R. (2011). Flying the flag, fleeing the state. *The New York Times*, April 24, A25.

23. Miller, D. D. & Sumaila, U. R. (2014). Flag use behavior and IUU activity within the international fishing fleet: Refining definitions and identifying areas of concern. *Marine Policy, 44*: 204–211.

24. EJF (Environmental Justice Foundation). (2005). *Pirates and Profiteers: How Pirate Fishing Fleets Are Robbing People and Oceans*. London, UK: Environmental Justice Foundation.

25. Hamad, H. B. (2015). Corporate social responsibility (CSR) in the shipping industry: A disturbing mechanism between maritime security needs and seafarers' welfare. *Studies of Organisational Management & Sustainability, 3*(1): 1–13.

26. Liddick, D. (2014). The dimensions of a transnational crime problem: The case of IUU fishing. *Trends in Organized Crime, 17*(4): 290–312.

27. Negret, C. F. L. (2016). Pretending to be Liberian and Panamanian; flags of convenience and the weakening of the nation state on the high seas. *Journal of Maritime Law and Commerce, 47*: 1–28.

28. Free economic zones are areas designated by countries, where both local and international companies can conduct business without being taxed or are taxed very lightly. Currently, 29 countries have such zones, with Italy offering 22 such zones, followed by United Arab Emirates (12) and Egypt (10).

29. Petrossian, G., Marteache, N. & Viollaz, J. (2014). Where do "undocumented" fish land? An empirical assessment of port characteristics for IUU fishing. *European Journal on Criminal Policy and Research, 21*(3): 337–351.

30. Marteache, N., Viollaz, J. & Petrossian, G. A. (2015). Factors influencing the choice of a safe haven for offloading illegally caught fish: A comparative analysis of developed and developing economies. *Crime Science, 4*(32). DOI: 10.1186/s40163-0045-2.

31. EJF (Environmental Justice Foundation). (2007). *Pirate Fish on Your Plate—Tracking Illegally-Caught Fish from West Africa into the European Market*. London, UK: Environmental Justice Foundation.

32. Ibid.

33. EJF (Environmental Justice Foundation). (2009). *Lowering the Flag—Ending the Use of Flags of Convenience by Pirate Fishing Vessels*. London, UK: Environmental Justice Foundation.

34. Gianni, M. & Simpson, W. (2005). *The Changing Nature of High Seas Fishing: How Flags of Convenience Provide Cover for Illegal, Unreported and Unregulated*

Fishing. Australian Government, Department of Agriculture, Fisheries and Forestry, International Transport Workers' Federation, and WWF International.

35. Ibid.

36. Kroodsma, D. A., Miller, N. A. & Roan, A. (2017). The global view of transshipment: Preliminary findings. *Global Fishing Watch and SkyTruth.* February 2017. https://globalfishingwatch.org/wp-content/uploads/GlobalViewOfTransshipment_Aug2017.pdf.

37. Ewell, C., Cullis-Suzuki, S., Ediger, M., Hocevar, J., Miller, D. & Jacquet, J. (2017). Potential ecological and social benefits of a moratorium on transshipment on the high seas. *Marine Policy, 81*: 293–300.

38. EJF (Environmental Justice Foundation). (2013). *Transshipment at Sea: The Need for a Ban in West Africa.* London, UK: Environmental Justice Foundation. 'http://ejfoundation.org/sites'default/files/public/Pirate%20Fishing%20Exposed.pdf.

39. Ewell, C., Cullis-Suzuki, S., Ediger, M., Hocevar, J., Miller, D. & Jacquet, J. (2017). Potential ecological and social benefits of a moratorium on transshipment on the high seas. *Marine Policy, 81*: 293–300.

40. FFA. (2016). *The Harmonized Minimum Terms and Conditions for Foreign Vessel Access.* Conservation Measure Amended by FFC99. http://www.apo-observers.org/docs/FFA_HMTC_2016.pdf.

41. European Commission. (2010). *EC Regulation 1005/2008 to Prevent, Deter and Eliminate Illegal, Unreported and Unregulated (IUU) Fishing.* Brussels: European Commission. https://eur-lex.europa.eu/legal-content/EN/ALL/?uri=celex%3A32008R1005.

42. WCPFC. (2006). *Western and Central Pacific Fisheries Commission Boarding and Inspection Procedures, Conservation and Management Measure 06-08.* Papeete, Tahiti. Western & Central Pacific Fisheries Commission. https://www.wcpfc.int/doc/cmm-2006-08/western-and-central-pacific-fisheries-commission-boarding-and-inspection-procedures.

43. SEAFO. (2006). *Conservation Measure 03/06 on an Interim Prohibition of Transshipments at Sea in the SEAFO Convention Area and to Regulate Transshipment in Port.* Swakopmund, Namibia: South East Atlantic Fisheries Organisation.

44. Ewell, C., Cullis-Suzuki, S., Ediger, M., Hocevar, J., Miller, D. & Jacquet, J. (2017). Potential ecological and social benefits of a moratorium on transshipment on the high seas. *Marine Policy, 81*: 293–300.

45. Bray, K. (Ed.). (2000). A global review of illegal, unreported and unregulated (IUU) fishing. One of the series of papers prepared as background documents for the Expert Consultation on Illegal, Unreported and Unregulated Fishing Organized by the Government of Australia in Cooperation with FAO, Sydney, Australia, May 15–19.

46. Asia-Pacific Economic Cooperation. (2008). *Assessment of Impacts of Illegal, Unreported and Unregulated Fishing in the Asia-Pacific.* Asia-Pacific Economic Cooperation Fisheries Working Group.

47. Sumaila, U. R., Alder, J. & Keith, H. (2006). Global scope and economics of illegal fishing. *Marine Policy, 30*(6): 696–703.

48. Mora, C., Myers, R. A., Coll, M., Libralato, S., Pitcher, T. J., Sumaila, R. U., Zeller, D., Watson, R., Gaston, K. J. & Worm, B., et al. (2009). Management effectiveness of the world's marine fisheries. *PLoS Biology, 7*(6): e1000131. DOI: 10.1371/journal.pbio.1000131.

49. Borsky, S. & Raschky, P. A. (2011). *A Spatial Econometric Analysis of Compliance with an International Environmental Agreement on Open Access Resources.* Monash

University. https://ageconsearch.umn.edu/record/124425/files/2012AC%20Raschky%20CP.pdf.

50. Martini, M. (2013). *Illegal, Unreported and Unregulated Fishing and Corruption*. U4 Anti-Corruption Resource Centre. www.U4.no.

51. Standing, A. (2008). *Corruption and Industrial Fishing in Africa*. U4 Anti-Corruption Resource Centre, U4 Brief. http://www.u4.no/publications/corruption-and-commercial-fisheries-in-africa/.

52. EJF (Environmental Justice Foundation). (2012). *Pirate Fishing Exposed: The Fight against Illegal Fishing in West Africa and the EU*. London, UK: Environmental Justice Foundation. https://ejfoundation.org/resources/downloads/Pirate-Fishing-Exposed.pdf.

53. Martini, M. (2013). *Illegal, Unreported and Unregulated Fishing and Corruption*. U4 Anti-Corruption Resource Centre. www.U4.no.

54. Ibid.

55. Hanich, Q. & Tsamenyi, M. (2009). Managing fisheries and corruption in the Pacific Islands region. *Marine Policy, 33*: 386–392.

56. Ibid.

57. Ibid.

58. Martini, M. (2013). *Illegal, Unreported and Unregulated Fishing and Corruption*. U4 Anti-Corruption Resource Centre. http://www.u4.no/publications/illegal-unreported-and-unregulated-fishing-and-corruption/.

59. Standing, A. (2008). *Corruption and Industrial Fishing in Africa*. U4 Anti-Corruption Resource Centre, U4 Brief. http://www.u4.no/publications/corruption-and-commercial-fisheries-in-africa/.

60. Ibid.

61. Ibid.

62. Ibid.

63. Standing, A. (2015). *Corruption and State-Corporate Crime in Fisheries*. U4 Issue. http://www.u4.no/publications/corruption-and-state-corporate-crime-in-fisheries/.

64. Sundström, A. (2013). Corruption in the commons: Why bribery hampers enforcement of environmental regulations in South African fisheries. *International Journal of the Commons, 7*(2): 454–472.

65. EJF (Environmental Justice Foundation). (2007). *Pirate Fish on Your Plate—Tracking Illegally-Caught Fish from West Africa into the European Market*. London, UK: Environmental Justice Foundation.

66. Asia-Pacific Economic Cooperation. (2008). *Case Study on Illegal, Unreported and Unregulated Fishing off the East Coast of Peninsular Malaysia*. Final Report submitted by Sea Resources Management Sdn Bhd. APEC Publication Number APEC#208-FS-01.4.

67. Palma, M. A. & Tsamenyi, M. (2008). *Case Study on the Impacts of Illegal, Unreported and Unregulated Fishing in the Sulawesi Sea*. APEC Fisheries Working Group. Australian National Centre for Ocean Resources and Security.

68. Gillett, R. & van Santen, G. (2008). *Optimizing Fisheries Benefits in the Pacific Islands: Major Issues and Constraints*. World Bank Group. Discussion Paper 44457. https://openknowledge.worldbank.org/bitstream/handle/10986/8078/444570ESW0whit1B0Fisheries01PUBLIC1.pdf?sequence=1.

69. Kuperan, K. & Sutinen, J. G. (1998). Blue water crime: Deterrence, legitimacy, and compliance in fisheries. *Law & Society Review, 32*(2): 309–337.

70. Nielsen, J. R. & Mathiesen, C. (2003). Important factors influencing rule compliance in fisheries: Lessons from Denmark. *Marine Policy, 27*: 409–416.

71. King, D. M., Price, E., Van Buren, A., Shearin, C., Mengerink, K., Porter, R. D., Sutinen, J. G., Rosenberg, A. & Swasey, J. H. (2010). *An Economic, Legal and Institutional Assessment of Enforcement and Compliance in Federally Managed U.S. Commercial Fisheries*. University of Maryland, Center for Environmental Science. Selected Reports and Publications.

72. Varkey, D. A., Ainsworth, C. H., Pitcher, T. J., Goram, Y. & Sumaila, R. (2010). Illegal, unreported and unregulated fisheries catch in Raja Ampat Regency, Eastern Indonesia.*Marine Policy, 34*: 228–236.

73. Ibid, p. 235.

Chapter 4

1. Wortley, R. & Mazerolle, L. (Eds.). (2008). *Environmental Criminology and Crime Analysis*. Portland, OR: Willan Publishing.

2. Cornish, D. B. & Clarke, R. V. G. (1986). *The Reasoning Criminal: Rational Choice Perspectives on Offending*. New York, NY: Springer-Verlag.

3. Wortley, R. & Mazerolle, L. (Eds.). (2008). *Environmental Criminology and Crime Analysis*. Portland, OR: Willan Publishing.

4. Cornish, D. B. & Clarke, R. V. G. (1986). *The Reasoning Criminal: Rational Choice Perspectives on Offending*. New York, NY: Springer-Verlag.

5. Felson, M. & Clarke, R. V. (1998). *Opportunity Makes the Thief: Practical Theory for Crime Prevention*. London, UK: Home Office, Policing and Reduction Crime Unit.

6. March, J. G. (1994). *A Primer on Decision Making: How Decisions Happen*. New York, NY: The Free Press, A Division of Simon and Schuster, Inc.

7. Wortley, R. & Mazerolle, L. (Eds.). (2008). *Environmental Criminology and Crime Analysis*. Portland, OR: Willan Publishing.

8. Clarke, R. V. (1980). "Situational" crime prevention: Theory and practice. *British Journal of Criminology, 20*(2): 136–147.

9. Clarke, R. V. (2008). Situational crime prevention. In Wortley, R. & Mazerolle, L. (Eds.). *Environmental Criminology and Crime Analysis*, p. 179. New York, NY: Routledge.

10. Ibid.

11. Cullen, F. T. & Agnew, R. (2006). *Criminological Theory: Past to Present*. Third Edition. Los Angeles, CA: Roxbury Publishing Company.

12. Clarke, R. V. (2008). Situational crime prevention. In Wortley, R. & Mazerolle, L. (Eds.). *Environmental Criminology and Crime Analysis,* p. 142. New York, NY: Routledge.

13. Brantingham, P. L. & Brantingham, P. J. (2005). Situational crime prevention as key component in embedded crime prevention. *Canadian Journal of Criminology and Criminal Justice, 47*(2): 271–292.

14. Clarke, R. V. (2008). Situational crime prevention. In Wortley, R. & Mazerolle, L. (Eds.). *Environmental Criminology and Crime Analysis*. New York, NY: Routledge.

15. Clarke, R. V. (Ed.). (1997). *Situational Crime Prevention: Successful Case Studies*. Second Edition. Monsey, NY: Criminal Justice Press, p. 20.

16. Brantingham, P. L. & Brantingham, P. J. (2005). Situational crime prevention as key component in embedded crime prevention. *Canadian Journal of Criminology and Criminal Justice, 47*(2): 271–292.

17. Wilson, J. Q. & Kelling, G. L. (March, 1982). Broken windows. *The Atlantic Monthly,* pp. 29–38.

18. Brantingham, P. L. & Brantingham, P. J. (2005). Situational crime prevention as key component in embedded crime prevention. *Canadian Journal of Criminology and Criminal Justice, 47*(2): 271–292.

19. Clarke, R. V. (2008). Situational crime prevention. In Wortley, R. & Mazerolle, L. (Eds.). *Environmental Criminology and Crime Analysis.* New York, NY: Routledge.

Chapter 5

1. EEZs are marine areas that have been established under the Law of the Sea and include sea zones over which states have special rights to use the marine resources. A country's EEZ has been established by the Law of the Sea to stretch from the seaward edge of the state's territorial sea to 200 nautical miles from its coast. When EEZs of the countries overlap, it is up to the states to decide their maritime boundaries. The general rule established by the Law of the Sea is that the point within the overlapping area defaulted to the nearest state.

2. Angola, Argentina, Australia, Bangladesh, Brazil, Canada, Chile, China, Denmark, Ecuador, Egypt, Faeroes, France, Germany, Ghana, Iceland, India, Indonesia, Iran, Ireland, Italy, Japan, Latvia, Malaysia, the Maldives, Mexico, Morocco, Myanmar, Namibia, New Zealand, Nigeria, North Korea, Norway, Pakistan, Peru, the Philippines, Poland, Portugal, Russia, Senegal, South Africa, South Korea, Spain, Sri Lanka, Sweden, Taiwan, Thailand, Turkey, Ukraine, the United Kingdom, the United States, Vietnam, Yemen.

3. FAO (Food and Agriculture Organization of the United Nations). (1998). Fishing operations: Vessel monitoring systems. *FAO Technical Guidelines for Responsible Fisheries*, No. 9, supra note 2, para 3.2.5. Rome.

4. Flewwelling, P. (2004). *An Introduction to Monitoring, Control and Surveillance Systems for Capture Fisheries.* FAO Fisheries Technical Paper 338. http://www.fao.org/DOCREP/003/V4250E/V4250E00.htm#toc.

5. FAO Code of Conduct. (1995). *Adopted at the 28th Session of the FAO Conference*. Rome, Italy. October 31.

6. Palma, M. A., Tsamenyi, M. & Edeson, W. (2010). *Promoting Sustainable Fisheries: The International Legal and Policy Framework to Combat Illegal, Unreported and Unregulated Fishing.* Leiden, Netherlands: Martinus Nijhoff Publishers.

7. Mees, C. (2000). *Points to Remember in Connection with Armed Boardings*. Technical Paper Presented at the Regional Workshop on Fisheries Monitoring, Control, and Surveillance, Kuala Lumpur and Kuala Terengganu, Malaysia, June 29–July 3. GCP/INT/648/NORI. Rome, FAO, at 147–150.

8. The PEW Environmental Group. (2010). *Port State Performance Report: Putting Illegal, Unreported and Unregulated Fishing on the Radar.* https://www.pewtrusts.org/~/media/post-launch-images/2015/04/2015_april_pew_port-state-performance—putting-iuu-on-radar(1).pdf.

9. Sea distance between countries is measured differently. First, there are set marine traffic routes that are used by ships to move in the seas. Second, this distance is measured

in nautical miles rather than miles, kilometers, or other measurements used in distance calculations. Taking these two factors into consideration, distances are calculated between each centrally located port along the coastline of the countries examined and all 10 POCs. These calculations were made through the online sea distance/voyage calculator available through www.ports.com.

10. For example, see Petrossian, G. A. (2015). Preventing illegal, unreported and unregulated (IUU) fishing: A situational approach. *Biological Conservation, 189*: 39–48.

11. These researchers examined these 54 important fishing countries only.

12. Pitcher, T., Kalikoski, D. & Pramod, G. (Eds.). (2006). Evaluations of compliance with the FAO (UN) code of conduct for responsible fisheries. *Fisheries Centre Research Report, 14*(2): 1191, The Fisheries Centre, UBC.

13. For the MCS measures, the scale of 0–10 indicated ranged from weakest (or no such measure in place) to strongest. For the "degree of illegal fishing," the score of "0" indicated no such activity recoded, "2.5," occasionally; "5," often; "7.5," a great deal-half as much as legal vessels; and "10," almost as much as, or more than, legal vessels. For the sake of clarity, these scores are converted into "0–2.5," low; "2.5–5," medium; "5–7.5," high; and "7.5–10," very high.

14. Sea Around Us Project. (2010). *Sea Around Us: A Ten-Year Retrospective, 1999–2009.* Fisheries Centre, The University of British Columbia, p. 2. Vancouver, Canada.

15. Wertheim, E. (2007). *Naval Institute Guide to Combat Fleets of the World: Their Ships, Aircraft, and Systems.* Fifteenth Edition. Annapolis, MD: U.S. Naval Institute Press.

16. These include the Maldives, Iran, India, Vietnam, Thailand, the Philippines, North Korea, Nigeria, Myanmar, Malaysia, Angola, Turkey, and Peru.

17. These include Turkey, Ecuador, Sri Lanka, Bangladesh, Egypt, Angola, North Korea, and Nigeria.

18. This was measured as the distance stretching 1,500 nautical miles from the port and the country's centrally located port.

19. See Petrossian, G. A. (2015). Preventing illegal, unreported and unregulated (IUU) fishing: A situational approach. *Biological Conservation, 189*: 39–48; Petrossian, G. A., Marteache, N. & Viollaz, J. (2015). Where do "undocumented" fish land? An empirical assessment of port characteristics for IUU fishing. *European Journal on Criminal Policy and Research, 21*(3): 337–351; Marteache, N., Viollaz, J. & Petrossian, G. A. (2015). Factors influencing the choice of a safe haven for offloading illegally caught fish: A comparative analysis of developed and developing economies. *Crime Science, 4*(1): 32.

20. FAO (Food and Agriculture Organization of the United Nations). (1998). Fishing operations: Vessel monitoring systems. *FAO Technical Guidelines for Responsible Fisheries,* No. 9, supra note 2, para 3.2.5. Rome.

Chapter 6

1. Epps, V. & Graham, L. (2011). *International Law: Examples and Explanations.* New York, NY: Aspen Publishers.

2. The "high seas" were first defined under the 1958 Convention as any part of the sea that was not either territorial sea or internal waters. As per Article 86 of the 1982 Convention, the high seas are defined as "all parts of the sea that are not included in the exclusive economic zone, in the territorial sea or the internal waters of a State, or in the archipelagic waters of an archipelagic State." Internal waters include lakes and rivers.

Bays, as well as waters that lie on the landward side of the baseline from which the territorial sea is measured, are also considered part of the state's internal waters. (See Epps, V. & Graham, L. (2011). *International Law: Examples and Explanations*. New York, NY: Aspen Publishers.)

3. Straddling stocks are fish stocks that occur in or migrate through more than one country's EEZ.

4. Palma, M. A., Tsamenyi, M. & Edeson, W. (2010). *Promoting Sustainable Fisheries: The International Legal and Policy Framework to Combat Illegal, Unreported and Unregulated Fishing*. Leiden, Netherlands: Martinus Nijhoff Publishers.

5. Interpol. (2018). *Projects*. https://www.interpol.int/Crime-areas/Environmental-crime/Projects/Project-Scale.

6. Poling, G. B. & Cronin, C. (2017). *Illegal, Unreported, and Unregulated Fishing as a National Security Threat*. Center for Strategic & International Studies. https://csis-prod.s3.amazonaws.com/s3fs-public/publication/171102_Poling_IUUFishing_Web.pdf?fxf_ZS98YbFth8SnVM242pH0VutBYw2v.

7. NOAA. (2017). *Improving International Fisheries Management*. Report to Congress Pursuant to Section 403(a) of the Magnuson-Stevens Fishery Conservation and Management Reauthorization Act of 2006. https://www.fisheries.noaa.gov/feature-story/improving-international-fisheries-management-highlights-2015-iuu-biennial-report.

8. Vulcan. (2017). *SkyLight Platform Aims to Combat Illegal Fishing*. http://www.vulcan.com/news/articles/2017/skylight-launch.

9. The Conversation. (2018). *How Blockchain Is Strengthening Tuna Traceability to Combat Illegal Fishing*. http://theconversation.com/how-blockchain-is-strengthening-tuna-traceability-to-combat-illegal-fishing-89965.

10. Fishcoin. (2018). *Fishcoin: A Blockchain Based Data Ecosystem for the Global Seafood Industry*. https://fishcoin.co/files/fishcoin.pdf.

11. Karapetyan, S. (September, 2018). *Project FishGuard: Seychelles to Monitor Illegal Fishing with Unmanned Drones*. Seychelles News Agency. http://www.seychellesnewsagency.com/articles/9754/Project+FishGuard+Seychelles+to+monitor+illegal+fishing+with+unmanned+drones.

12. Beurteaux, D. (April 9, 2018). Turning birds into spies against illegal fishing. *Hakai Magazine*. https://www.hakaimagazine.com/news/turning-birds-into-spies-against-illegal-fishing/.

13. WildLeaks. (2015). *Manta Slaughter in the Philippines*. https://wildleaks.org/manta-slaughter-in-the-philippines/.

14. EJF (Environmental Justice Foundation). (2005). *Pirates and Profiteers: How Pirate Fishing Fleets Are Robbing People and Oceans*. London, UK: Environmental Justice Foundation.

15. This term is used to refer to illegal catches.

16. Green, M. (February 24, 2016). The black fish: Undercover with the vigilantes fighting organized crime at sea. *The Guardian*. https://www.theguardian.com/environment/2016/feb/24/black-fish-undercover-with-vigilantes-fighting-organised-crime-at-sea.

17. Njock, J. & Allison, E. H. (2008). Institutional innovations in fisheries co-management. In Westlund, L., Holvoet, K. & Kébé, M. (Eds.). *Achieving Poverty Reduction through Responsible Fisheries: Lessons from West and Central Africa*. FAO Fisheries and Aquaculture Technical Paper, No. 513. Rome: FAO.

18. Ibid.

19. Ibid.

20. See Fishery Solutions Center at http://fisherysolutionscenter.edf.org for a map and full descriptions of over 200 rights-based management (RBM) programs implemented worldwide. The database also provides interactive maps of the programs by type, species targeted, gear used, and geographic region.

21. Halpern, B. S. (2003). The impact of marine reserves: Do reserves work and does reserve size matter? *Ecological Applications, 13*(1) Supplement: 117–137.

22. Kura, Y., Revenga, C., Hoshino, E. & Mock, G. (2004). *Fishing for Answers: Making Sense of the Global Fish Crisis*. Washington, DC: World Resource Institute.

23. Oceana. (2009). *Fisheries Subsidies, Overcapacity and Overfishing*. https://usa.oceana.org/sites/default/files/Fisheries_Subsidies_Backgrounder_REVISED_Oct09_FINAL_0.pdf.

24. Undercurrent News. (December 18, 2015). *WTO Members Agree to Reduce Fishing Subsidies*. https://www.undercurrentnews.com/2015/12/18/wto-members-agree-to-reduce-fishing-subsidies/.

25. FAO (Food and Agriculture Organization of the United Nations). (2010). FAO Fisheries and Aquaculture Report No. 932. United Nations. FIRF/R932 (En).

26. Sheikh, P. A. & Corn, M. L. (2005). *The Convention on International Trade in Endangered Species of Wild Fauna and Flora (CITES): Background and issues*. Congressional Research Service, Library of Congress. Order Code: RL32751.

27. Ibid.

28. Sheikh, P. A. & Corn, M. L. (2016). *The Convention on International Trade in Endangered Species of Wild Fauna and Flora (CITES)*. Congressional Research Service 7-5700. RL32751. https://fas.org/sgp/crs/misc/RL32751.pdf.

29. Ibid.

30. Ibid.

31. A comprehensive list and short descriptions of U.S. fisheries laws can be found at http://www.hg.org/fisheries-law.html. A thorough review of fisheries laws in select European, Asian, African, and Latin American countries, as well as in North America, New Zealand, and the South Pacific, is provided in a legislative study prepared for the Food and Agriculture Organization of the United Nations by Cacaud, P., Kuruc, M. & Spreij, M. (2003). *Sanctions in Fisheries Law*. Legislative Study No. 82. Food and Agriculture Organization of the United Nations. Rome, Italy.

32. Poling, G. B. & Cronin, C. (2017). *Illegal, Unreported, and Unregulated Fishing as a National Security Threat*. Center for Strategic & International Studies. https://csis-prod.s3.amazonaws.com/s3fs-public/publication/171102_Poling_IUUFishing_Web.pdf?fxf_ZS98YbFth8SnVM242pH0VutBYw2v.

33. Palma, M. A., Tsamenyi, M. & Edeson, W. (2010). *Promoting Sustainable Fisheries: The International Legal and Policy Framework to Combat Illegal, Unreported and Unregulated Fishing*. Leiden, Netherlands: Martinus Nijhoff Publishers.

34. Ibid.

35. Mitolidis, S. (2016). *The European Union Action towards the International Fight against IUU Fishing*. http://www.seafoodsummit.org/wp-content/uploads/2016/03/Future%20of%20Anti-IUU-Mitolidis.pdf.

36. The EU-IUU Regulation. (n.d.). *The EU-IUU Regulation: Building on Success: EU Progress in the Global Fight against Illegal Fishing*. https://ejfoundation.org/

reports/the-eu-iuu-regulation-building-on-success-eu-progress-in-the-global-fight-against-illegal-fishing.

37. Palma, M. A., Tsamenyi, M. & Edeson, W. (2010). *Promoting Sustainable Fisheries: The International Legal and Policy Framework to Combat Illegal, Unreported and Unregulated Fishing*. Leiden, Netherlands: Martinus Nijhoff Publishers, p. 188.

38. Deere, C. (1999). *Eco-labelling and Sustainable Fisheries*. Rome: FAO.

39. These data are as of 2018. For more on the Ecolabel Index, see http://www.ecolabelindex.com.

Chapter 7

1. In this section, we will provide a general overview of analytical strategies. For more in-depth discussion on the strategies mentioned in this section, see Moreto, W. D. & Pires, S. F. (2018). *Wildlife Crime: An Environmental Criminology and Crime Science Perspective*. Durham, NC: Carolina Academic Press.

2. This acronym stands for Concealable, Removable, Available, Valuable, Enjoyable, and Disposable. See Chapter 2 for a discussion of these species.

3. This acronym stands for concealable, available, processable, transferrable, useable, removable, enjoyable, desirable. For an in-depth discussion of the concept, see Moreto, W. D. & Lemieux, A. M. (2015). From CRAVED to CAPTURED: Introducing a product-based framework to examine illegal wildlife markets. *European Journal on Criminal Policy and Research*, *21*(3): 303–320.

4. Cornish, D. B. (1994). The procedural analysis of offending and its relevance for situational prevention. *Crime Prevention Studies, 3*: 151–196.

5. Moreto, W. D. & Clarke, R. V. (2011). Script analysis of the transnational illegal market in endangered species: Dream and reality. In Leclerc, B. & Wortley, R. (Eds.). *Cognition and Crime: Offender Decision-Making and Script Analyses*. New York, NY: Routledge.

6. Petrossian, G. A. & Pezzella, F. S. (2018). IUU fishing and seafood fraud: Using crime script analysis to inform intervention. *ANNALS, AAPSS, 679*: 121–139.

7. Interpol. (2018). International law enforcement cooperation in the fisheries sector: A guide for law enforcement practitioners. https://stopillegalfishing.com/news-articles/global-fisheries-enforcement-guide-published-by-interpol/.

8. Brantingham, P. L. & Brantingham, P. J. (1993). Environment, routine, and situation: Toward a pattern theory of crime. In Clarke, R. & Felson, M. (Eds.). *Routine Activity and Rational Choice: Advances in Criminological Theory, Volume 5*. New Brunswick, NJ: Transaction Publishers.

9. Petrossian, G. A. (2018). A micro-spatial analysis of opportunities for IUU fishing in 23 Western African countries. *Biological Conservation, 225*: 31–41.

10. Pala, A., Zhang, J., Zhuang, J. & Allen, N. (2018). Behavior analysis of illegal fishing in the Gulf of Mexico. *Journal of Homeland Security and Emergency Management, 15*(1): 1–10.

11. Cohen, L. E. & Felson, M. (1979). Social change and crime rate trends: A routine activity approach. *American Sociological Review, 44*: 588–608.

12. Clarke, R. V. (1999). *Hot Products: Understanding, Anticipating and Reducing Demand for Stolen Goods*. Police Research Series, Paper 112. Policing and Reducing Crime Unit, Research Development and Statistics Directorate. London, UK: Home Office.

13. Pires, S. F. & Clarke, R. V. (2012). Are parrots CRAVED? An analysis of parrot poaching in Mexico. *Journal of Research in Crime and Delinquency, 49*(1): 122–146.

14. Petrossian, G. A. & Clarke, R. V. (2015). Explaining and controlling illegal commercial fishing: An application of the CRAVED theft model. *British Journal of Criminology, 54*(1): 73–90.

15. Pires, S. F. & Petrossian, G. A. (2016). Understanding parrot trafficking between illicit markets in Bolivia: An application of the CRAVED model. *International Journal of Comparative and Applied Criminal Justice, 40*(1): 63–77.

16. This framework uses the *concealable, removable, available*, and *enjoyable* elements of the CRAVED model; replaces *valuable* and *disposable* with *desirable* and *transferable*; and adds *processable* and *usable* elements.

17. Moreto, W. D. & Lemieux, A. M. (2015). From CRAVED to CAPTURED: Introducing a product-based framework to examine illegal wildlife markets. *European Journal on Criminal Policy and Research, 21*: 303–320.

18. Petrossian, G. A. & Clarke, R. V. (2014). Explaining and controlling illegal commercial fishing: An application of the CRAVED theft model. *British Journal of Criminology, 54*(1): 73–90.

19. Petrossian, G., Weis, J. S. & Pires, S. F. (2015). Factors affecting crab and lobster species subject to IUU fishing. *Ocean & Coastal Management, 106*: 29–34.

20. Clarke, R. V. & Eck, J. E. (2007). *Understanding Risky Facilities*. U.S. Department of Justice, Office of Community Oriented Policing. Problem Oriented Guides for Police, Problem Solving Tolls Series, No. 6; see also Eck, J. E., Clarke, R. V. & Guerette, R. T. (2007). Risky facilities: Crime concentration in homogeneous sets of establishments and facilities. In Farrell, G., Bowers, K. J., Johnson, S. D. & Townsley, M. (Eds.). *Imagination for Crime Prevention: Crime Prevention Studies, Volume 17*. Monsey, NY: Criminal Justice Press.

21. Petrossian, G. A., Marteache, N. & Viollaz, J. (2015). Where do "undocumented" fish land? An empirical assessment of port characteristics for IUU fishing. *European Journal on Criminal Policy and Research, 21*(3): 337–351.

22. Weekers, D. P. & Zahnow, R. (2018). Risky facilities: Analysis of illegal recreational fishing in the Great Barrier Reef Marine Park, Australia. *Australian & New Zealand Journal of Criminology*. https://doi.org/10.1177/0004865818804021.

23. Rossmo, D. K. (2000). *Geographic Profiling*. Boca Raton, FL: CRC Press.

24. Eckblom, P. (2005). Designing products against crime. In N. Tilley (Ed.). *Handbook of Crime Prevention and Community Safety*. Portland, OR: Willan Publishing, pp. 203–244.

25. Freiberg, A. (1997). Regulating markets for stolen property. *Australian and New Zealand Journal of Criminology, 30*(3): 237–258. For a discussion on how this strategy can be used to reduce the illegal wildlife trade, see Schneider, J. (2008). Reducing the illicit trade in endangered wildlife: The market reduction approach. *Journal of Contemporary Criminal Justice, 24*(3): 274–295; and Schneider, J. (2012). *Sold into Extinction: The Global Trade in Endangered Species*. Santa Barbara, CA: Praeger.

26. See, for example, Guerette, R. T. & Bowers, K. J. (2009). Assessing the extent of crime displacement and diffusion of benefits: A review of situational crime prevention evaluations. *Criminology, 47*(4): 1331–1368. The authors evaluated 102 crime-prevention projects to determine whether these interventions resulted in displacement and diffusion of benefits. The researchers observed that displacement happened only

26 percent of the time, and in 27 percent of the cases, there was diffusion of benefits. Moreover, in 13 studies, while geographic displacement did occur, it was much smaller than the effect of the intervention in the targeted area, suggesting that the intervention was still effective.

27. Bowers, K. & Johnson, S. (2003). Measuring the geographical displacement and diffusion of benefit effects of crime prevention activity. *Journal of Quantitative Criminology, 19*(3): 275–301.

28. UNODC (United Nations Office on Drugs and Crime). (2011). *Transnational Organized Crime in the Fishing Industry: Focus on Trafficking in Persons, Smuggling of Migrants and Illicit Drugs Trafficking*. Vienna: United Nations.

29. Asia-Pacific Economic Cooperation. (2008). *Case Study on Illegal, Unreported and Unregulated Fishing Off the East Coast of Peninsular Malaysia.* Final report submitted by Sea Resources Management Sdn Bhd. APEC Publication Number: APEC#208-FS-01.4.

30. Interpol. (2018). International law enforcement cooperation in the fisheries sector: A guide for law enforcement practitioners. https://stopillegalfishing.com/news-articles/global-fisheries-enforcement-guide-published-by-interpol/.

31. Skaret, G. & Pitcher, T. J. (2006). *An Estimation of Compliance of the Fisheries of Norway with Article 7 (Fisheries Management) of the UN Code of Conduct for Responsible Fishing.* University of British Columbia, Fisheries Center Research Reports.

32. Pramod, G. & Pitcher, T. J. (2006). *An Estimation of Compliance of the Fisheries of Namibia with Article 7 (Fisheries Management) of the UN Code of Conduct for Responsible Fishing.* University of British Columbia, Fisheries Center Research Reports.

33. Varkey, D. & Pitcher, T. J. (2006). *An Estimation of Compliance of the Fisheries of Iceland with Article 7 (Fisheries Management) of the UN Code of Conduct for Responsible Fishing.* University of British Columbia, Fisheries Center Research Reports (p. 21).

Bibliography

Adovasio, J. M., Soffer, O. & Klima, B. (1996). Upper Palaeolithic fiber technology: Interlaced woven finds from Pavlov I, Czech Republic, c. 26,000 years ago. *Antiquity, 70*(269): 526–534.

Africa Progress Panel. (2014). *Launch Remarks—Kofi Annan.* Geneva: Africa Progress Panel Report. https://www.afdb.org/fileadmin/uploads/afdb/Documents/Project-and-Operations/Africa_Progress_Report_2014.PDF.

Agnew, D. J. & Barnes, C. T. (2004). Economic aspects and drivers of IUU fishing: Building a framework in OECD. In *Fish Piracy. Combating Illegal, Unreported and Unregulated Fishing*. Paris: OECD Publishing.

Agnew, D. J., Pearce, J., Pramod, G., Peatman, T., Watson, R., Beddington, J. R. & Pitcher, T. J. (2009). Estimating the worldwide extent of illegal fishing. *PLoS ONE, 4*(2): e4570. DOI: 10.1371/journal.pone.0004570.

Alverson, D. L., Freeberg, M. K., Murawski, S. A. & Pope, J. G. (1994). *A Global Assessment of Fisheries Bycatch and Discards.* FAO Fisheries Technical Paper No. 339. Rome, Italy: FAO.

Alward, G. L. (1932). *The Sea Fisheries of Great Britain and Ireland: A Record of the Development of the Fishing Industry and Its Worldwide Ramifications. Notes from the Author's Diary, 1854–1928, with Charts, Photographs and Illustrations.* Grismby, UK: Albert Gait. Cited in Pitcher, T. J. & Lam, M. E. (2015). Fish commoditization and the historical origins of catching fish for profit. *Maritime Studies, 14*(2): 1–19.

Anon. (2002). *Assessment Report on Fisheries and Fisheries Related Species and Habitat Issues.* Bergen, Norway: North Sea Conference. Cited in Gelchu, A. & Pauly, D. (2007). Growth and distribution of port-based global fishing effort within countries' EEZs from 1970 to 1995. *Fisheries Centre Research Reports, 15*(4): 1–86.

APO (Asian Productivity Organization). (1988). *Fishing Industry in Asia and Pacific.* Hong Kong, China: Medical International Limited.

Asada, Y., Hirasawa, Y. & Nagasaki, F. (1983). *Fisheries Management in Japan.* FAO Fisheries Technical Paper No. 238. Rome, Italy: FAO.

Asia-Pacific Economic Cooperation. (2008). *Assessment of Impacts of Illegal, Unreported and Unregulated Fishing in the Asia-Pacific.* Asia-Pacific Economic Cooperation Fisheries Working Group.

Asia-Pacific Economic Cooperation. (2008). *Case Study on Illegal, Unreported and Unregulated Fishing off the East Coast of Peninsular Malaysia.* Final Report submitted by Sea Resources Management Sdn Bhd. APEC Publication Number: APEC#208-FS-01.4.

Bailey, C. (1997). Lessons from Indonesia's 1980 trawler ban. *Marine Policy, 21*(3): 225–235.

Behnam, A. & Faust, P. (2003). Twilight of flag state control. *Ocean Yearbook Online, 17* (1). DOI: https://doi.org/10.1163/221160003X00096.

Beurteaux, D. (April 9, 2018). Turning birds into spies against illegal fishing. *Hakai Magazine.* https://www.hakaimagazine.com/news/turning-birds-into-spies-against-illegal-fishing/.

Black, R. (2007). Action urged on illegal fishing. *BBC News.* Accessed October 21, 2013, http://news.bbc.co.uk/2/hi/6602639.stm.

Bockstoce, J. (1986). *Whales, Ice, and Men: The History of Whaling in the Western Arctic.* Seattle, WA: University of Washington Press.

Bonini, S., Saran, N. & Stein, L. (2011). *Design for Sustainable Fisheries—Modeling Fishery Economics.* https://www.mckinsey.com/~/media/mckinsey/industries/social%20sector/our%20insights/saving%20our%20resources%20a%20design%20for%20sustainable%20fisheries/design-for-sustainable-fisheries-modeling-fishery-economics.ashx.

Borsky, S. & Raschky, P. A. (2011). *A Spatial Econometric Analysis of Compliance with an International Environmental Agreement on Open Access Resources.* Monash University. https://ageconsearch.umn.edu/record/124425/files/2012AC%20Raschky%20CP.pdf.

Bowers, K. & Johnson, S. (2003). Measuring the geographical displacement and diffusion of benefit effects of crime prevention activity. *Journal of Quantitative Criminology, 19*(3): 275–301.

Brandt, A. V. (1972). *Fish Catching Methods of the World. Revised and Enlarged.* London, UK: Fishing News (Books) Ltd.

Brantingham, P. L. & Brantingham, P. J. (1993). Environment, routine, and situation: Toward a pattern theory of crime. In Clarke, R. & Felson, M. (Eds.). *Routine Activity and Rational Choice: Advances in Criminological Theory, Volume 5.* New Brunswick, NJ: Transaction Publishers.

Brantingham, P. L. & Brantingham, P. J. (2005). Situational crime prevention as key component in embedded crime prevention. *Canadian Journal of Criminology and Criminal Justice, 47*(2): 271–292.

Braund, S. M. (Trans.). (2004). *Juvenal and Persius: Loeb Classical Library 91.* Cambridge, MA: Harvard University Press. Cited in Pitcher, T. J. & Lam, M. E. (2015). Fish commoditization and the historical origins of catching fish for profit. *Maritime Studies, 14*(2): 1–19.

Bray, K. (Ed.). (2000). A global review of illegal, unreported and unregulated (IUU) fishing. One of the series of papers prepared as background documents for the Expert

Consultation on Illegal, Unreported and Unregulated Fishing Organized by the Government of Australia in Cooperation with FAO, Sydney, Australia, May 15–19.

Brewer, D. J. & Friedman, R. F. (1990). *Fish and Fishing in Ancient Egypt*. Cairo, Egypt: The American University in Cairo Press.

Brian, R. (1985). Recent developments in Australian fishing industry. In Cleveland, B. C. (Ed.). *Adjustments to Changes in Fisheries Law and Economics*. FAO Fisheries Technical Paper No. 269, Suppl. Rome, Italy: FAO, pp. 21–26.

Brothers, N. P. (1991). Albatross mortality and associated bait loss in the Japanese longline fishery in the southern ocean. *Biological Conservation, 55*: 255–268.

Bryant, D., Burke, L., McManus, J. & Spalding, M. (1998). *Reefs at Risk: A Map-based Indicator of Threats to the World's Coral Reefs*. World Resources Institute, International Center for Living Aquatic Resources Management, World Conservation Monitoring Centre, United Nations Environment Programme, 56 p. https://wriorg.s3.amazonaws.com/s3fs-public/pdf/reefs.pdf?_ga=2.57804330.1209804127.1548434425-2003236149.1548434425.

Cacaud, P., Kuruc, M. & Spreij, M. (2003). Administrative *Sanctions in Fisheries Law*. Legislative Study No. 82. Food and Agriculture Organization of the United Nations. Rome, Italy.

Caldwell, R. & Fox, H. (2006). Recovery from blast fishing on coral reefs: A tale of two scales. *Ecological Applications, 16*(5): 1631–1635.

Campbell, S. K. & Butler, V. L. (2010). Archaeological evidence for resilience of Pacific Northwest salmon populations and the socioecological system over the last ~7500 years. *Ecology and Society, 15*(1): 17.

Chen, T. (2012). Easy pickings for abalone smugglers. *Wall Street Journal, China Real Time*. http://blogs.wsj.com/chinarealtime/2012/03/30/easy-pickings-for-abalone-smugglers/.

Chevallier, R. (2017). Safeguarding Tanzania's coral reefs: The case of illegal blast fishing. *South African Institute of International Affairs, Policy Insights, 46*. Johannesburg, South Africa.

Chidambaram, K. (1963). Development of mechanized fishing fleets in the Indo-Pacific council region. In Nui, L. B. & Jan-Old Trrmg, J.-O. (Eds.). *Indo-Pacific Fisheries Council Proceedings*. Section II, Symposium Papers. Bangkok: IPFC Secretariat, FAO regional office for Asia and the Far East, pp. 2–27.

Clark, M. R. & Koslow, J. A. (2007). Impacts of fisheries on seamounts. In Pitcher, T. J., Morato, T., Hart, P. J. B., Clark, M. R., Haggan, N. & Santos, R. S. (Eds.). *Seamounts: Ecology, Fisheries, and Conservation*. Fish and Aquatic Resources Series, 12. Oxford, UK: Blackwell Publishing.

Clarke, R. V. (1980). "Situational" crime prevention: Theory and practice. *British Journal of Criminology, 20*(2): 136–147.

Clarke, R. V. (Ed.). (1997). *Situational Crime Prevention: Successful Case Studies*. Second Edition. Monsey, NY: Criminal Justice Press, p. 20.

Clarke, R. V. (1999). *Hot Products: Understanding, Anticipating, and Reducing Demand for Stolen Goods*. Police Research Series, Paper 112. Policing and Reducing Crime Unit, Research Development and Statistics Directorate. London, UK: Home Office.

Clarke, R. V. (2008). Situational crime prevention. In Wortley, R. & Mazerolle, L. (Eds.). *Environmental Criminology and Crime Analysis*. New York, NY: Routledge, p. 179.

Clarke, R. V. & Cornish, D. B. (2001). Rational choice. In Paternoster, R. & Bachman, R. (Eds.). *Explaining Criminals and Crime*. Los Angeles, CA: Roxbury Publishing Company, p. 24.

Clarke, R. V. & Eck, J. E. (2007). *Understanding Risky Facilities*. Washington, DC: U.S. Department of Justice, Office of Community Oriented Policing Services. Problem Oriented Guides for Police, Problem Solving Tools Series, No. 6; see also Eck, J. E., Clarke, R. V. & Guerette, R. T. (2007). Risky facilities: Crime concentration in homogeneous sets of establishments and facilities. In Farrell, G., Bowers, K. J., Johnson, S. D. & Townsley, M. (Eds.). *Imagination for Crime Prevention. Crime Prevention Studies, Volume 17*. Monsey, NY: Criminal Justice Press.

Clover, C. (2004). *The End of the Line: How Overfishing Is Changing the World and What We Eat*. London, UK: Ebury Press, Random House, p. 129.

Cohen, L. E. & Felson, M. (1979). Social change and crime rate trends: A routine activity approach. *American Sociological Review, 44*: 588–608.

Commission for the Conservation of Antarctic Marine Living Resources (CCAMLR). (2004). *Report of the Twenty-Third Meeting of the Commission*. Hobart, Australia, October 25–November 4. CCAMLR-XXIII, Tasmania, Australia, at 31.

Concoran, T. H. (1963). Roman fish sauces. *The Classical Journal, 58*(5): 204–210.

The Conversation. (2018). *How Blockchain Is Strengthening Tuna Traceability to Combat Illegal Fishing*. http://theconversation.com/how-blockchain-is-strengthening-tuna-traceability-to-combat-illegal-fishing-89965.

Cornish, A. S. & McKellar, D. (1998). A history of fishing with explosives and poisons in Hong Kong waters. *Naga, The ICLARM Quarterly*, July–September, 4–9.

Cornish, D. B. (1994). The procedural analysis of offending and its relevance for situational prevention. *Crime Prevention Studies*, 3: 151–196.

Cornish, D. B. & Clarke, R. V. G. (1986). *The Reasoning Criminal: Rational Choice Perspectives on Offending*. New York, NY: Springer-Verlag.

Cullen, F. T. & Agnew, R. (2006). *Criminological Theory: Past to Present*. Third Edition. Los Angeles, CA: Roxbury Publishing Company.

Curtis, R. I. (2005). Sources for production and trade of Greek and Roman processed fish. In Bekker-Nielsen, T. (Ed.). *Ancient Fishing and Fish Processing in the Black Sea Region*. Esbjerg, Denmark: Aarhus University Press.

Cushing, D. H. (1988). *The Provident Ocean*. London, UK: Cambridge University Press.

Dalzell, P., Adams, T. & Polunin, N. (1997). Coastal fisheries in the Pacific Islands. *Oceanographic Literature Review, 44*(5): 516.

Daniels, A., Gutiérrez, M., Fanjul, G., Guereña, A., Matheson, I. & Watkins, K. (2016). *Western Africa's Missing Fish: The Impacts of Illegal, Unreported and Unregulated Fishing and Under-Reporting Catches by Foreign Fleets*. London, UK: Overseas Development Institute.

Deere, C. (1999). *Eco-labelling and Sustainable Fisheries*. Rome, Italy: FAO.

Doughty, R. W. & Carmichael, V. (2011). *The Albatross and the Fish: Linked Lives in the Open Seas*. Austin, TX: University of Texas Press.

Duplés-Agier, H. (1852). Ordonnances inédites de Philippe le Bel et de Philippe le Long Sur La Police de la Pêche Fluviale. *Bibliothèque de l'Ecole des Chartes, 14*: 49. Cited in Hoffman, R. C. (1996). Economic development and aquatic ecosystems in medieval Europe. *The American Historical Review, 101*(3): 631–669.

Eckblom, P. (2005). Designing products against crime. In Tilley, N. (Ed.). *Handbook of Crime Prevention and Community Safety*. Portland, OR: Willan Publishing, pp. 203–244.

The Economist. (July 7, 2018). *Avast, Me Hearties: If Drones Ruled the Waves*. https://www.economist.com/the-world-if/2018/07/07/avast-me-hearties.

EJF (Environmental Justice Foundation). (2005). *Pirates and Profiteers: How Pirate Fishing Fleets Are Robbing People and Oceans*. London, UK: Environmental Justice Foundation.

EJF (Environmental Justice Foundation). (2007). *Illegal Driftnetting in the Mediterranean*. London, UK: Environmental Justice Foundation.

EJF (Environmental Justice Foundation). (2007). *Pirate Fish on Your Plate-Tracking Illegally-Caught Fish from West Africa into the European Market*. London, UK: Environmental Justice Foundation.

EJF (Environmental Justice Foundation). (2009). *Lowering the Flag—Ending the Use of Flags of Convenience by Pirate Fishing Vessels*. London, UK: Environmental Justice Foundation.

EJF (Environmental Justice Foundation). (2010). *All at Sea—The Abuse of Human Rights Aboard Illegal Fishing Vessels*. London, UK: Environmental Justice Foundation.

EJF (Environmental Justice Foundation). (2012). *Pirate Fishing Exposed: The Fight against Illegal Fishing in West Africa and the EU*. London, UK: Environmental Justice Foundation. https://ejfoundation.org/resources/downloads/Pirate-Fishing-Exposed.pdf.

EJF (Environmental Justice Foundation). (2013). *Transshipment at Sea: The Need for a Ban in West Africa*. London, UK: Environmental Justice Foundation. https://ejfoundation.org/resources/downloads/Pirate-Fishing-Exposed.pdf.

Engler, M. & Parry-Jones, R. (2007). *Opportunity or Threat: The Role of the European Union in Global Wildlife Trade*. Brussels: TRAFFIC Europe.

Epps, V. & Graham, L. (2011). *International Law: Examples and Explanations*. New York, NY: Aspen Publishers.

The EU-IUU Regulation. (n.d.). *The EU-IUU Regulation: Building on Success: EU Progress in the Global Fight against Illegal Fishing*. https://ejfoundation.org/reports/the-eu-iuu-regulation-building-on-success-eu-progress-in-the-global-fight-against-illegal-fishing.

European Commission. (2010). *EC Regulation 1005/2008 to Prevent, Deter and Eliminate Illegal, Unreported and Unregulated (IUU) Fishing*. Brussels: European Commission. https://eur-lex.europa.eu/legal-content/EN/ALL/?uri=celex%3A32008R1005.

European Parliament. (2014). *Illegal, Unreported and Unregulated Fishing: Sanctions in the EU*. Strasbourg: European Parliament. http://www.europarl.europa.eu/RegData/etudes/STUD/2014/529069/IPOL_STU(2014)529069_EN.pdf.

Ewell, C., Cullis-Suzuki, S., Ediger, M., Hocevar, J., Miller, D. & Jacquet, J. (2017). Potential ecological and social benefits of a moratorium on transshipment on the high seas. *Marine Policy, 81*: 293–300.

Falaye, A. E. (2008). Illegal Unreported Unregulated (IUU) Fishing in West Africa (Nigeria & Ghana). London, UK: *Marine Resources Assessment Group Ltd.*

FAO (Food and Agriculture Organization of the United Nations). (1998). Fishing operations: Vessel monitoring systems. *FAO Technical Guidelines for Responsible Fisheries,* No. 9, supra note 2, para 3.2.5. Rome, Italy.

FAO (Food and Agriculture Organization of the United Nations). (2007). *FAO Workshop on Vulnerable Ecosystems and Destructive Fishing in Deep-Sea Fisheries*. FAO Report No. 829. Rome, Italy.

FAO (Food and Agriculture Organization of the United Nations). (2010). *FAO Fisheries and Aquaculture Report No. 932*. United Nations. FIRF/R932 (En).

FAO (Food and Agriculture Organization of the United Nations). (2014). *FAO Yearbook. Fisheries and Aquaculture Statistics*. United Nations. Rome, Italy.

FAO (Food and Agriculture Organization of the United Nations). (2016). *FAO Fisheries and Aquaculture Report No. 932*. United Nations. FIRF/R932 (En).

FAO (Food and Agriculture Organization of the United Nations). (2016). *The State of World Fisheries and Aquaculture. Contributing to Food Security and Nutrition for All.* Rome, Italy: FAO.

FAO (Food and Agriculture Organization of the United Nations). (n.d.). Fisheries section: Glossary. *Fishing Industry*. www.fao.org/fi/glossary/default.asp.

FAO Code of Conduct. (1995). *Adopted at the 28th Session of the FAO Conference*. Rome, Italy. October 31.

Felson, M. & Clarke, R. V. (1998). *Opportunity Makes the Thief: Practical Theory for Crime Prevention*. London, UK: Home Office, Policing and Reduction Crime Unit.

Ferretti, F., Myers, R. A., Serena, F. & Lotze, H. K. (2008). Loss of large predatory sharks from the Mediterranean Sea. *Conservation Biology, 22*(4): 952–964.

FFA. (2016). *The Harmonized Minimum Terms and Conditions for Foreign Vessel Access.* Conservation Measure Amended by FFC99. http://www.apo-observers.org/docs/FFA_HMTC_2016.pdf.

Fishcoin. (2018). *Fishcoin: A Blockchain Based Data Ecosystem for the Global Seafood Industry*. https://fishcoin.co/files/fishcoin.pdf.

Fishery Solutions Center. (2018). *Design Stories: Spanish Galicia Goose Barnacle Cofradía System*. http://fisherysolutionscenter.edf.org/design-stories/spanish-galicia-goose-barnacle-cofrad%C3%ADa-system.

Fishing Vessel types. Trawlers. Technology Fact Sheets. In: *FAO Fisheries and Aquaculture Department* [online]. Rome, Italy. Updated May 9, 2001. [Cited October 15, 2013]. http://www.fao.org/fishery/vesseltype/10/en.

Fleming, A. (January 16, 2018). Pulse fishing: Is it OK to "electrocute" fish? *BBC News*. http://www.bbc.com/news/world-europe-42692924.

Flewwelling, P. (2004). *An Introduction to Monitoring, Control and Surveillances Systems for Capture Fisheries*. FAO Fisheries Technical Paper No. 338. http://www.fao.org/DOCREP/003/V4250E/V4250E00.htm#toc.

Freiberg, A. (1997). Regulating markets for stolen property. *Australian and New Zealand Journal of Criminology, 30*(3): 237–258.

Freitas, B. (2015). *Illegal Fishing: Which Fish Species Are at Highest Risk from Illegal and Unreported Fishing?* Washington, DC: World Wildlife Fund.

Garcia, S. M. & Newton, C. (1997). Current situation, trends and prospects in world capture fisheries. In Pikitch, E. L., Huppert, D. D., & Sissenwine, M. P. (Eds.). *Global Trends: Fisheries Management*, pp. 3–27. Seattle, WA: American Fisheries Society Symposium 20, June 14–16.

Gelchu, A. & Pauly, D. (2007). Growth and distribution of port-based global fishing effort within countries' EEZs from 1970 to 1995. *Fisheries Centre Research Reports, 15* (4): 1–86.

George, R. (2011). Flying the flag, fleeing the state. *The New York Times*, April 24, A25.

Gianni, M. & Simpson, W. (2005). *The Changing Nature of High Seas Fishing: How Flags of Convenience Provide Cover for Illegal, Unreported and Unregulated Fishing*. Australian Government, Department of Agriculture, Fisheries and Forestry, International Transport Workers' Federation, and WWF International.

Gillett, R. & van Santen, G. (2008). *Optimizing Fisheries Benefits in the Pacific Islands: Major Issues and Constrains*. Washington, DC: World Bank Group. Discussion Paper 44457. https://openknowledge.worldbank.org/bitstream/handle/10986/8078/444570ESW0whit1B0Fisheries01PUBLIC1.pdf?sequence=1.

Gilman, E. (2004). *References on Seabird Bycatch in Longline Fisheries*. Honolulu, HI: Blue Ocean Institute.

Gravitz, M. (2014). *Plundering the Seas: The Damage from Pirate Fishing on US Fishermen & Communities*. Seattle, WA: Marine Conservation Institute. http://www.marine-conservation.org/media/filer_public/filer_public/2014/09/04/mci_iuu_report.pdf.

Green, E. K. (2011). Telephoning fish: An examination of the creative deviance used by wildlife violators in the United States. *International Journal of Rural Criminology, 1*(1): 23–39.

Green, M. (February 24, 2016). The black fish: Undercover with the vigilantes fighting organized crime at sea. *The Guardian*. https://www.theguardian.com/environment/2016/feb/24/black-fish-undercover-with-vigilantes-fighting-organised-crime-at-sea.

Greenpeace. (2015). *How the Illegal Totoaba Trade in Hong Kong Is Pushing Mexico's Vaquita to the Brink of Extinction*. https://www.greenpeace.org/eastasia/Global/eastasia/publications/campaigns/Oceans/HK%20Totoaba%20Trade_Greenpeace%20Media%20Briefing.pdf.

Guerette, R. T. & Bowers, K. J. (2009). Assessing the extent of crime displacement and diffusion of benefits: A review of situational crime prevention evaluations. *Criminology, 47*(4): 1331–1368.

Gullard, J. A. (1974). *The Management of Marine Fisheries*. Bristol, England: Scientechnica (Publishers) Ltd.

Halpern, B. S. (2003). The impact of marine reserves: Do reserves work and does reserve size matter? *Ecological Applications, 13*(1) Supplement: 117–137.

Hamad, H. B. (2015). Corporate social responsibility (CSR) in the shipping industry: A disturbing mechanism between maritime security needs and seafarers' welfare. *Studies of Organisational Management & Sustainability, 3*(1): 1–13.

Hanich, Q. & Tsamenyi, M. (2009). Managing fisheries and corruption in the Pacific Island region. *Marine Policy, 33*: 386–392.

High Seas Task Force. (2006). *Closing the Net: Stopping Illegal Fishing on the High Seas*. Governments of Australia, Canada, Chile, Namibia, New Zealand, and the United Kingdom, WWF, IUCN, and the Earth Institute at Columbia University.

Hilborn, R. & Hilborn, U. (2012). *Overfishing: What Everyone Needs to Know*. New York, NY: Oxford University Press.

Hoffman, R. C. (1996). Economic development and aquatic ecosystems in medieval Europe. *The American Historical Review, 101*(3): 631–669.

Hofford, A. (2010). *Yet More Endangered Species on Sale in Hong Kong, This Time the "Larke Croaker" Fish*. Accessed December 2016, http://www.alexhoffordphotography.com/node/2303.

House of Ocean. (2015). *IUU: Is It a Bird or a Plane? Is It Illegal Fishing, Unregulated Fishing or Crime? Look to the Fish Stocks Agreement for Answers*. Accessed October 6, 2017, https://houseofocean.org/2015/10/03/iuu-is-it-a-bird-or-a-plane-is-it-illegal-fishing-unregulated-fishing-or-crime-look-to-the-fish-stocks-agreement-for-answers/.

Hoydal, K. (2004). IUU fishing in NEAFC: How big is the problem and what have we done? *Fish Piracy: Combating Illegal, Unreported and Unregulated Fishing*. Paris: OECD, p. 161.

International Monitoring, Control and Surveillance Network. (2014). *Timor Leste Wins First-Ever STOP IUU Fishing Award*. http://gfetw.org/wp-content/uploads/2013/08/Press-Release_Feb-20_English.pdf.

Interpol. (2018). *International Law Enforcement Cooperation in the Fisheries Sector: A Guide for Law Enforcement Practitioners*. https://stopillegalfishing.com/news-articles/global-fisheries-enforcement-guide-published-by-interpol/.

ITF (International Transport Workers' Federation). (2006). *Out of Sight, Out of Mind: Seafarers, Fishers & Human Rights*. http://www.itfseafarers.org/files/extranet/-1/2259/HumanRights.pdf.

Jackson, J. B., Kirby, M. X., Berger, W. H., Bjorndal, K. A., Botsford, L. M., Bourque, B. J., . . . Hughes, T. P. (2001). Historical overfishing and the recent collapse of coastal ecosystems. *Science, 293*(5530): 629–637.

Jia, J. & Chen, J. (2001). *Sea Farming and Sea Ranching in China*. FAO Fisheries Technical Paper No. 418. Rome, Italy: FAO.

Johnson, V. (1992). Artisanal fishermen and intermediaries in Mozambique. In Tvedten, I. & Hersoug, B. (Eds.). *Fishing for Development: Small-Scale Fisheries in Africa*. Sweden: Nordiska Afrikainstitutet (Scandinavian Institute of African Studies), pp. 70–90.

Kacsynski, V. M. & Fluharty, D. L. (2002). European policies in West Africa: Who benefits from fisheries agreements? *Marine Policy, 26*: 75–93.

Karapetyan, S. (September, 2018). *Project FishGuard: Seychelles to Monitor Illegal Fishing with Unmanned Drones*. Seychelles News Agency. http://www.seychellesnewsagency.com/articles/9754/Project+FishGuard+Seychelles+to+monitor+illegal+fishing+with+unmanned+drones.

King, D. M., Price, E., Van Buren, A., Shearin, C., Mengerink, K., Porter, R. D., Sutinen, J. G., Rosenberg, A. & Swasey, J. H. (2010). *An Economic, Legal and Institutional Assessment of Enforcement and Compliance in Federally Managed U.S. Commercial Fisheries*. College Park, MD: University of Maryland, Center for Environmental Science. Selected Reports and Publications.

Kroodsma, D. A., Miller, N. A. & Roan, A. (February, 2017). The global view of transshipment: Preliminary findings. *Global Fishing Watch and SkyTruth*. https://globalfishingwatch.org/wp-content/uploads/GlobalViewOfTransshipment_Aug2017.pdf.

Kuper, J. (2008.). *Oceans in Crisis. Situation Status*. https://stopillegalfishing.com/news-articles/oceans-in-crisis-by-jo-kuper-greenpeace-international-5/.

Kuperan, K. & Sutinen, J. G. (1998). Blue water crime: Deterrence, legitimacy, and compliance in fisheries. *Law & Society Review, 32*(2): 309–337.

Kura, Y., Revenga, C., Hoshino, E. & Mock, G. (2004). *Fishing for Answers: Making Sense of the Global Fish Crisis*. Washington, DC: World Resource Institute.

Kurlansky, M. (1997). *Cod: A Biography of the Fish That Changed the World.* New York, NY: Penguin Books.

Kurlansky, M. (2002). *Salt: A World History.* New York, NY: Walker and Company.

Lah, K. & Moya, A. (2016). "Aquatic cocaine": Fish bladders are latest Mexican smuggling commodity. *CNN Health.* http://www.cnn.com/2016/05/23/health/aquatic-cocaine-totoaba-bass-smuggling/index.html.

Lawson, R. M. & Kwei, E. R. (1974). *African Entrepreneurship and Economic Growth: A Case Study of the Fishing Industry of Ghana.* Accra: Ghana Universities Press.

Liddick, D. (2014). The dimensions of a transnational crime problem: The case of IUU fishing. *Trends in Organized Crime, 17*: 290–312.

Liddick, D. R. (2011). *Crimes against Nature: Illegal Industries and the Global Environment.* Santa Barbara, CA: Praeger.

Llácer, F. J. M. (2003). Open registers: Past, present and future. *Marine Policy, 27*(6): 513–523.

Longo, S. B. & Clark, B. (2012). The commodification of bluefin tuna: The historical transformation of the Mediterranean fishery. *Journal of Agrarian Change, 12*(2–3): 204–226.

Love, P. (2010). *Fisheries: While Stocks Last.* Paris: Organization for Economic Cooperation and Development.

MacSweet, I. M. (1983). *Markets for Fish and Fishery Products in Eastern Europe.* FAO Fisheries Technical Paper No. 241 FIIU/T241. Rome, Italy: FAO.

Mak, K. K. W., Yanase, H. & Renneberg, R. (2005). Cyanide fishing and cyanide detection in coral reef fish using chemical tests and biosensors. *Biosensors and Bioelectronics, 20*(12): 2581–2593.

March, J. G. (1994). *A Primer on Decision Making: How Decisions Happen.* New York, NY: The Free Press, A Division of Simon and Schuster, Inc.

Marine Conservation Society. (n.d.). *How Fish Are Caught.* https://www.mcsuk.org/media/seafood/Fishing_Methods.pdf.

Marteache, N., Viollaz, J. & Petrossian, G. A. (2015). Factors influencing the choice of a safe haven for offloading illegally caught fish: A comparative analysis of developed and developing economies. *Crime Science, 4*(1): 32. DOI: 10.1186/s40163-0045-2.

Martini, M. (2013). *Illegal, Unreported and Unregulated Fishing and Corruption.* U4 Anti-Corruption Resource Centre. www.U4.no.

McLeod, W. J. (1964). The flags-of-convenience problem. *South Carolina Law Review, 16*: 409–418.

Mees, C. (2000). *Points to Remember in Connection with Armed Boardings.* Technical Paper Presented at the Regional Workshop on Fisheries Monitoring, Control, and Surveillance, Kuala Lumpur and Kuala Terengganu, Malaysia, June 29–July 3. GCP/INT/648/NORI. Rome, Italy: FAO, pp. 147–150.

Miller, D. D. & Sumaila, U. R. (2014). Flag use behavior and IUU activity within the international fishing fleet: Refining definitions and identifying areas of concern. *Marine Policy, 44*: 204–211.

Mitolidis, S. (2016). *The European Union Action towards the International Fight against IUU Fishing.* http://www.seafoodsummit.org/wp-content/uploads/2016/03/Future%20of%20Anti-IUU-Mitolidis.pdf.

Monterey Bay Aquarium. (2013). *Fishing and Farming Methods.* Accessed October 15, 2013, https://www.seafoodwatch.org/ocean-issues/fishing-and-farming-methods.

Mora, C., Myers, R. A., Coll, M., Libralato, S., Pitcher, T. J., Sumaila, R. U., Zeller, D., Watson, R., Gaston, K. J. & Worm, B. (2009). Management effectiveness of the world's marine fisheries. *PLoS Biology, 7*(6): e1000131. DOI: 10.1371/journal.pbio.1000131.

Moreto, W. D. & Clarke, R. V. (2011). Script analysis of the transnational illegal market in endangered species: Dream and reality. In Leclerc, B. & Wortley, R. (Eds.). *Cognition and Crime: Offender Decision-making and Script Analyses.* New York, NY: Routledge.

Moreto, W. D. & Lemieux, A. M. (2015). From CRAVED to CAPTURED: Introducing a product-based framework to examine illegal wildlife markets. *European Journal on Criminal Policy and Research, 21*: 303–320.

Moreto, W. D. & Pires, S. F. (2018). *Wildlife Crime: An Environmental Criminology and Crime Science Perspective.* Durham, NC: Carolina Academic Press.

Morgan, G. R. & Staples, D. J. (2006). *The History of Industrial Marine Fisheries in Southeast Asia.* Bangkok, Thailand: Food and Agriculture Organization of the United Nations Regional Office for Asia and the Pacific.

Mosbergen, D. (January 4, 2017). U.S. Navy dolphins join the fight to save endangered vaquita porpoises. *Huffington Post.* https://www.huffingtonpost.com/entry/dolphins-vaquita-porpoise_us_586c9901e4b0d9a5945d2801.

MRAG. (2005). *Review of Impacts of Illegal, Unreported and Unregulated Fishing on Developing Countries.* Final Report. Report prepared for the UK's Department for International Development with support from the Norwegian Agency for Development Cooperation.

MRAG. (2008). *Study and Analysis of the Status of IUU Fishing in the SADC Region and an Estimate of the Economic, Social and Biological Impacts.* Volume 2-Main Report. https://mrag.co.uk/services/illegal-unreported-and-unregulated-fishing?page=1.

NEAFC. (2009). *Scheme of Control and Enforcement.* Article 29.

Negret, C. F. L. (2016). Pretending to be Liberian and Panamanian; flags of convenience and the weakening of the nation state on the high seas. *Journal of Maritime Law and Commerce, 47*: 1–28.

Neutel, A. M., Heesterbeek, J. A. P. & de Ruiter, P. C. (2002). Stability in real food webs: Weak links in long loops. *Science, 296*: 1120–1123.

Neuwinger, H. D. (2004). Plants used for poison fishing in tropical Africa. *Toxicon, 44*(4): 417–430.

Nielsen, J. R. & Mathiesen, C. (2003). Important factors influencing rule compliance in fisheries: Lessons from Denmark. *Marine Policy, 27*: 409–416.

Njock, J. & Allison, E. H. (2008). Institutional innovations in fisheries co-management. In Westlund, L., Holvoet, K., & Kébé, M. (Eds.). *Achieving Poverty Reduction through Responsible Fisheries: Lessons from West and Central Africa.* FAO Fisheries and Aquaculture Technical Paper No. 513. Rome, Italy: FAO.

Njofonjou, O. & Njock, J. C. (2000). African fisheries: Major trends and diagnostic of the 20th century. In *Microbehavior and Macroresults: Proceedings of the Tenth Biennial Conference of the International Institute for Fisheries Economics and Trade*, July 10–14. Corvallis, OR: Oregon State University, Compiled by Richard S. Johnston & Ann L. Shriver. International Institute of Fisheries Economics and Trade (IIFET), Corvallis, 2001.

NOAA. (2017). *Improving International Fisheries Management.* Report to Congress Pursuant to Section 403(a) of the Magnuson-Stevens Fishery Conservation and Management Reauthorization Act of 2006. https://www.fisheries.noaa.gov/feature-story/improving-international-fisheries-management-highlights-2015-iuu-biennial-report.

Oceana. (2009). *Fisheries Subsidies, Overcapacity and Overfishing.* https://usa.oceana.org/sites/default/files/Fisheries_Subsidies_Backgrounder_REVISED_Oct09_FINAL_0.pdf.

Oceana. (2017). *Fishing the Boundaries of Law: How the Exclusivity Clause in EU Fisheries Agreements Was Undermined.* http://usa.oceana.org/publications/reports/fishing-boundaries-law-how-exclusivity-clause-eu-fisheries-agreements-was.

Oceans, Ltd. (1990). *Reproductive Success in Atlantic Cod (Gadus morhua): The Potential Impact of Trawling.* A Report Prepared for the Newfoundland Inshore Fisheries Association.

Osieke, E. (1979). Flags of convenience vessels: Recent developments. *The American Journal of International Law, 73*: 604–627.

Österblom, H., Cunstable, A. & Fukumi, S. (2011). Illegal fishing and the organized crime analogy. *Trends in Ecology and Evolution, 26*(6): 261–262.

Pala, A., Zhang, J., Zhuang, J. & Allen, N. (2018). Behavior analysis of illegal fishing in the Gulf of Mexico. *Journal of Homeland Security and Emergency Management, 15*(1): 1–10.

Palma, M. A., Tsamenyi, M. & Edeson, W. (2010). *Promoting Sustainable Fisheries: The International Legal and Policy Framework to Combat Illegal, Unreported and Unregulated Fishing.* Leiden, Netherlands: Martinus Nijhoff Publishers.

Palma, M. A. & Tsamenyi, M. (2008). *Case Study on the Impacts of Illegal, Unreported and Unregulated Fishing in the Sulawesi Sea.* Wollongong, Australia: APEC Fisheries Working Group, Australian National Centre for Ocean Resources and Security.

Paul, L. (1994). *High Seas Driftnetting: The Plunder of the Global Commons.* http://earthtrust.org/archive/dnpaper/contents.html, as cited in EJF (Environmental Justice Foundation). (2007). *Illegal Driftnetting in the Mediterranean.* London, UK: Environmental Justice Foundation. ISBN No. 1-904523-11-0.

Pauly, D., Christensen, F., Dalsgaard, J., Froese, R. & Torres, F. (1998). Fishing down marine food webs. *Science, 279*(5352): 860–863.

Pauly, D., Watson, R. & Alder, J. (2005). Global trends in world fisheries: Impacts on marine ecosystems and food security. *Philosophical Transactions of the Royal Society, 360*: 5–12.

Petrossian, G. A. (2015). Preventing illegal, unreported and unregulated (IUU) fishing: A situational approach. *Biological Conservation, 189*: 39–48.

Petrossian, G. A. (2018). A micro-spatial analysis of opportunities for IUU fishing in 23 Western African countries. *Biological Conservation, 225*: 31–41.

Petrossian, G. A. & Clarke, R. V. (2014). Explaining and controlling illegal commercial fishing: An application of the CRAVED theft model. *British Journal of Criminology, 54*(1): 73–90.

Petrossian, G. A., Marteache, N. & Viollaz, J. (2015). Where do "undocumented" fish land? An empirical assessment of port characteristics for IUU fishing. *European Journal on Criminal Policy and Research, 21*(3): 337–351.

Petrossian, G. A. & Pezzella, F. S. (2018). IUU fishing and seafood fraud: Using crime script analysis to inform intervention. *ANNALS, AAPSS, 679*: 121–139.

Petrossian, G., Weis, J. S. & Pires, S. F. (2015). Factors affecting crab and lobster species subject to IUU fishing. *Ocean & Coastal Management, 106*: 29–34.

Pew Charitable Trusts. (December 10, 2013). *How to End Illegal Fishing*. PEW. https://www.pewtrusts.org/en/research-and-analysis/reports/2013/12/10/how-to-end-illegal-fishing.

The Pew Environmental Group. (2010). *Port State Performance: Putting Illegal, Unreported and Unregulated Fishing on the Radar*. https://www.pewtrusts.org/~/media/post-launch-images/2015/04/2015_april_pew_port-state-performance—putting-iuu-on-radar(1).pdf.

Phelps Bondaroff, T. N., Reitano, T. & Wietse, W. (2015). *The Illegal Fishing and Organized Crime Nexus: Illegal Fishing as Transnational Organized Crime*. The Global Initiative against Transnational Crime and the Black Fish. https://globalinitiative.net/wp-content/uploads/2015/04/the-illegal-fishing-and-organised-crime-nexus-1.pdf.

Pires, S. F. & Clarke, R. V. (2012). Are parrots CRAVED? An analysis of parrot poaching in Mexico. *Journal of Research in Crime and Delinquency, 49*(1): 122–146.

Pires, S. F. & Petrossian, G. A. (2016). Understanding parrot trafficking between illicit markets in Bolivia: An application of the CRAVED model. *International Journal of Comparative and Applied Criminal Justice, 40*(1): 63–77.

Pitcher, T., Kalikoski, D. & Pramod, G. (Eds.). (2006). Evaluations of compliance with the FAO (UN) code of conduct for responsible fisheries. *Fisheries Centre Research Reports, 14*(2): 1191, The Fisheries Centre, UBC.

Pitcher, T. J. & Lam, M. E. (2015). Fish commoditization and the historical origins of catching fish for profit. *Maritime Studies, 14*(2): 1–19.

Poling, G. B. & Cronin, C. (2017). *Illegal, Unreported and Unregulated Fishing as a National Security Threat*. Center for Strategic and International Studies in cooperation with National Geographic Pristine Seas. https://www.csis.org/analysis/illegal-unreported-and-unregulated-fishing-national-security-threat.

Pramod, G., Nakamura, K., Pitcher, T. J. & Delagran, L. (2014). Estimates of illegal and unreported fish in seafood imports to the USA. *Marine Policy, 48*: 102–113.

Pramod, G. & Pitcher, T. J. (2006). *An Estimation of Compliance of the Fisheries of Namibia with Article 7 (Fisheries Management) of the UN Code of Conduct for Responsible Fishing*. Vancouver, Canada: University of British Columbia, Fisheries Center Research Reports.

Putt, J. & Nelson, D. (n.d.). Crime in the Australian fishing industry. *Trends and Issues in Crime and Criminal Justice, 366*. https://aic.gov.au/publications/tandi/tandi366.

Radcliffe, W. (1921). *Fishing from the Earliest Times*. London, UK: John Murray.

Report on the Status of Groundfish Stocks in the Canadian Northwest Atlantic. (June, 1994). Atlantic Stock Assessment Secretariat, Department of Fisheries and Oceans, p. 19.

Roberts, C. (2007). *The Unnatural History of the Sea*. Washington, DC: Island Press.

Rossmo, D. K. (2000). *Geographic Profiling*. Boca Raton, FL: CRC Press.

Sahrhage, D. & Lundbeck, J. (1992). *A History of Fishing*. Berlin, Heidelberg: Springer-Verlag.

Schneider, J. (2008). Reducing the illicit trade in endangered wildlife: The market reduction approach. *Journal of Contemporary Criminal Justice, 24*(3): 274–295.

Schneider, J. (2012). *Sold into Extinction: The Global Trade in Endangered Species.* Santa Barbara, CA: Praeger.

Scott, C. (June 5, 2018). *Saving Real Fish with 3D Printed Robotic Fish.* https://3dprint.com/215846/3d-printed-robotic-fish-real-fish/.

Sea Around Us Project. (2010). *Sea Around Us: A Ten-Year Retrospective, 1999–2009.* Vancouver, Canada: Fisheries Centre, The University of British Columbia, p. 2.

SEAFO. (2006). *Conservation Measure 03/06 on an Interim Prohibition of Transshipments at Sea in the SEAFO Convention Area and to Regulate Transshipment in Port.* Swakopmund, Namibia: South East Atlantic Fisheries Organisation.

Sea Shepherd. (2018). *We Are Sea Shepherd.* https://seashepherd.org.

Sebater, M. (2007). *Prehistoric, Historic and Current Reef Fish Utilization and the Consequence of Human Population Growth on Fish Density, Biomass and Community Composition in American Samoa.* Bio. Rep. Series No. 2007-01. American Samoa: Department of Marine and Wildlife Resources.

Shaughnessy, T. & Tobin, E. (2006). Flags of inconvenience: Freedom and insecurity on the high seas. *Journal of International Law, 5*(1): 1–31.

Shaw, M. (2014). Transnational organized crime in Africa. In Albanese, J. S. & Reichel, P. L. (Eds.). *Transnational Organized Crime: An Overview from Six Continents.* Los Angeles, CA: Sage Publications, pp. 93–115.

Sheikh, P. A. & Corn, M. L. (2005). *The Convention on International Trade in Endangered Species of Wild Fauna and Flora (CITES): Background and Issues.* Congressional Research Service, Library of Congress. Order Code: RL32751. Washington, DC.

Sheikh, P. A. & Corn, M. L. (2016). *The Convention on International Trade in Endangered Species of Wild Fauna and Flora (CITES).* Congressional Research Service 7-5700. RL32751. https://fas.org/sgp/crs/misc/RL32751.pdf.

Skaret, G. & Pitcher, T. J. (2006). *An Estimation of Compliance of the Fisheries of Norway with Article 7 (Fisheries Management) of the UN Code of Conduct for Responsible Fishing.* Vancouver, Canada: University of British Columbia, Fisheries Center Research Report.

Solecki, J. J. (1979). A review of USSR fishing industry. *Ocean Management, 5*: 97–123.

Standing, A. (2008). *Corruption and Industrial Fishing in Africa.* Bergen, Norway: U4 Anti-Corruption Resource Centre. https://www.cmi.no/publications/file/3188-corruption-andindustrial-fishing-in-africa.pdf.

Stephens, D. W. & Krebs, J. R. (1986). *Foraging Theory.* Princeton, NJ: Princeton University Press.

Stiles, M. L., Kagan, A., Shaftel, E. & Lowell, B. (2013). *Stolen Seafood: The Impact of Pirate Fishing on Our Oceans.* Oceana. http://oceana.org/sites/default/files/reports/Oceana_StolenSeafood.pdf.

Sumaila, U. R., Alder, J. & Keith, H. (2006). Global scope and economics of illegal fishing. *Marine Policy, 30*(6): 696–703.

Sundstrom, A. (2013). Corruption in the commons: Why bribery hampers enforcement of environmental regulations in South African fisheries. *International Journal of the Commons, 7*(2): 454–472.

Swann, J. (2002). *Fishing Vessels Operating Under Open Registers and the Exercise of Flag State Responsibilities: Information and Options*. FAO Fisheries Circular No. 980, UN FAO, Rome, Italy.

Swartz, W., Sumaila, U. R., Watson, R. & Pauly, D. (2010). Sourcing seafood for the three major markets: The EU, Japan and the USA. *Marine Policy, 34*: 1366–1373.

Takayama, S. (1963). Mechanization of fishing fleets in Japan. In Nui, L. B. & Jan-Old Trrmg, J.-O. (Eds.). *Indo-Pacific Fisheries Council Proceedings*. 10th Session, Seoul, Republic of Korea, Section II Symposium Papers, pp. 1–117. Bangkok: IPFC Secretariat, FAO Regional Office for Asia and the Far East, pp. 1–117.

Telesetsky, A. (2015). Laundering fish in the global undercurrents: Illegal, unreported, and unregulated fishing and transnational organized crime. *Ecology Law Quarterly, 41* (4): 939–996.

Thiele, W. (1999). *Global Trends in Fishing Technology and Their Effects on Fishing Power and Capacity*. Busan, Republic of Korea: FAO Regional Workshop on the Effects of Globalization and Deregulation of Marine Capture Fisheries in Asia.

TRAFFIC Bulletin. (October 29, 2014). *Wildlife Watch: Strengthening Capacities for the Protection of Nature in Mexico*. Accessed November 29, 2016, http://www.traffic.org/home/2014/10/29/wildlife-watch-strengthening-capacities-for-the-protection-o.html.

Tvedten, I. & Hersoug, B. (Eds.). (1992). *Fishing for Development: Small-Scale Fisheries in Africa*. Sweden: Nordiska Afrikainstituten (The Scandinavian Institute of African Studies).

Undercurrent News. (December 18, 2015). *WTO Members Agree to Reduce Fishing Subsidies*. https://www.undercurrentnews.com/2015/12/18/wto-members-agree-to-reduce-fishing-subsidies/.

UNEP. (1994). *Staff Training Materials for the Management of Marine Protected Areas*. RCU/EAS Technical Report Series No. 4, 216–220.

United Nations. (2005). *Convention on the High Seas*. Done at Geneva on April 29, 1958. Entered into force on September 30, 1962. *Treaty Series,* Vol. 450.

United Nations. (2008). Conference of Parties to the United Nations Convention against Transnational Organized Crime, Report of the Conference of Parties to the United Nations Convention against Transnational Organized Crime on Its Fourth Session, Vienna, October 8–17. CTOC/COP/2008/19, December 1, 2008, para. 210.

United Nations Fisheries and Aquaculture Department. (2015). *FAO Fishery Statistics, FAO FishStat Database*. http://www.fao.org/fishery/statistics/en.

United States, House of Representatives, Armed Services Committee Hearing, Vessel Operations Under Flags of Convenience and their Implications on National Security. (2002). http://commdocs.house.gov/committees/security/has164220.000/has164220_0.HTM.

UNODC (United Nations Office on Drugs and Crime). (2011). *Transnational Organized Crime in the Fishing Industry: Focus on Trafficking in Persons, Smuggling of Migrants and Illicit Drugs Trafficking*. Vienna: United Nations.

Urbina, I. (July 27, 2015). Sea slaves: The human misery that feeds pets and livestock. *The New York Times*.

Van Uhm, D. P. (2016). *Uncovering the Illegal Wildlife Trade: Inside the World of Poachers, Smugglers and Traders*. New York, NY: Springer.

Vanuatu Maritime Service Limited. (2016). *Vanuatu Maritime Ships—Vanuatu International Ship Registry*. New York, NY. http://vanuatumaritimeships.com.

VaquitaCPR. (2017). *How Does the Acoustic Detection of the Vaquita Work?* https://www.vaquitacpr.org/how-does-the-acoustic-detection-of-the-vaquita-work/.

Varkey, D. A., Ainsworth, C. H., Pitcher, T. J, Goram, Y. & Sumaila, R. (2010). Illegal, unreported and unregulated fisheries catch in Raja Ampat Regency, Eastern Indonesia. *Marine Policy, 34*: 228–236.

Varkey, D. & Pitcher, T. J. (2006). *An Estimation of Compliance of the Fisheries of Iceland with Article 7 (Fisheries Management) of the UN Code of Conduct for Responsible Fishing*. Vancouver, Canada: University of British Columbia, Fisheries Center Research Report, p. 21.

Volkov, V. (2002). *Violent Entrepreneurs. The Use of Force in the Making of Russian Capitalism*. Ithaca, NY and London, UK: Cornell University Press.

Vulcan. (2017). *SkyLight Platform Aims to Combat Illegal Fishing*. http://www.vulcan.com/news/articles/2017/skylight-launch.

Warchol, G. & Harrington, M. (2016). Exploring the dynamics of South Africa's illegal abalone trade via routine activities theory. *Trends in Organized Crime, 19*: 21–41.

Watson, R., Hoshino, E., Beblow, J., Revenga, C., Kura, Y. & Kitchingman, A. (2004). Fishing gear associated with global marine catches. *Fisheries Centre Research Reports, 12*(6): 3–32.

WCPFC. (2006). *Western and Central Pacific Fisheries Commission Boarding and Inspection Procedures, Conservation and Management Measure 06–08*. Papeete, Tahiti. Western & Central Pacific Fisheries Commission. https://www.wcpfc.int/doc/cmm-2006-08/western-and-central-pacific-fisheries-commission-boarding-and-inspection-procedures.

Weekers, D. P. & Zahnow, R. (2018). Risky facilities: Analysis of illegal recreational fishing in the Great Barrier Reef Marine Park, Australia. *Australian & New Zealand Journal of Criminology*. https://doi.org/10.1177/0004865818804021.

Wertheim, E. (2007). *Naval Institute Guide to Combat Fleets of the World: Their Ships, Aircraft and Systems*. Fifteenth Edition. Annapolis, MD: Naval Institute Press.

White, A. T., Vogt, H. P. & Arin, T. (2000). The Philippine coral reefs under threat: The economic losses caused by reef destruction. *Marine Pollution Bulletin, 40*: 598–605.

WildLeaks. (2015). *Manta Slaughter in the Philippines*. https://wildleaks.org/manta-slaughter-in-the-philippines/.

Williamson, M. J. (2013). *New Report Highlights Recreational Fishing's Broad Economic and Conservation Impact*. American Sportfishing Association Press Release. Accessed October 7, 2013. https://asafishing.org/new-report-highlights-recreational-fishings-broad-economic-and-conservation/.

Wilson, J. Q. & Kelling, G. L. (March, 1982). Broken windows. *The Atlantic Monthly*, pp. 29–38.

Wimpenny, R. S. (1953). *The Plaice: Being the Buckland Lectures for 1949*. London, UK: Edward Arnold Publishers.

Wortley, R. & Mazerolle, L. (Eds.). (2008). *Environmental Criminology and Crime Analysis*. Portland, OR: Willan Publishing.

Wyler, L. S. & Sheick, P. A. (2013). *International Illegal Trade in Wildlife: Threats and US Policy*. Washington, DC: Congressional Research Service Report to Congress.

Yagoub, M. (2016). *"Narco Fish" Trafficking Easy Money for Mexico Criminals.* Accessed November 2016, https://www.insightcrime.org/news/brief/narco-fish-trafficking-easy-money-for-mexico-criminals.

Yirka, B. (February 24, 2012). *Robot Fish Found Able to Lead Real Fish.* https://phys.org/news/2012-02-robot-fish-real-video.html.

Yumiko, K., Revenga, C., Hoshino, E. & Mock, G. (2004). *Fishing for Answers: Making Sense of the Global Fish Crisis.* Washington, DC: World Resources Institute. http://www.wri.org/sites/default/files/pdf/fishanswer_fulltext.pdf.

Zabyelina, Y. (2014). The "fishy business": A qualitative analysis of the illicit market in Black Caviar. *Trends in Organized Crime, 17*(3): 181–198.

Index

Page numbers followed by an italicized *t* or *f* refer to a table or a figure, respectively.

About the Author

GOHAR A. PETROSSIAN, PhD, is an assistant professor in the Department of Criminal Justice at CUNY—John Jay College of Criminal Justice, New York. She has published in top criminal justice, social science, and conservation journals, including *The British Journal of Criminology*, *The Annals of the American Academy of Political and Social Sciences*, *Biological Conservation, Ocean & Coastal Management,* and *Oryx*. Petrossian holds a doctorate in criminal justice from Rutgers University–Newark, School of Criminal Justice.

www.ingramcontent.com/pod-product-compliance
Ingram Content Group UK Ltd.
Pitfield, Milton Keynes, MK11 3LW, UK
UKHW020843290926
15470UKWH00004B/1525